MINIATURE SUCCULENT PLANTS

PLANTS
Wonders of Nature

Timothy Polnaszek, PhD

TAN Books
Gastonia, North Carolina

Plants: Wonders of Nature © 2021 Timothy Polnaszek

All rights reserved. With the exception of short excerpts used in critical review, no part of this work may be reproduced, transmitted, or stored in any form whatsoever, without the prior written permission of the publisher.

Unless otherwise noted, Scripture quotations are from the Revised Standard Version of the Bible—Second Catholic Edition (Ignatius Edition), copyright © 2006 National Council of the Churches of Christ in the United States of America. Used by permission. All rights reserved.

Excerpts from the English translation of the Catechism of the Catholic Church for use in the United States of America © 1994, United States Catholic Conference, Inc.—Libreria Editrice Vaticana. Used with permission.

Cover & interior design and typesetting by www.davidferrisdesign.com

ISBN: 978-1-5051-1771-4

Published in the United States by
TAN Books
PO Box 269
Gastonia, NC 28053

www.TANBooks.com

Printed in the United States of America

"Where were you when I laid the foundation of the earth?"

—Job 38:4

CONTENTS

Preface VIII

Introduction XI

Chapter 1: What Is a Plant, Anyway? 1

Chapter 2: Plants' Superpower: Generating Food from Light 9

Chapter 3: Bryophytes: Living the Simple Life 19

Chapter 4: Ferns and Friends: Vascular, Seedless Plants 27

Chapter 5: Gymnosperms: Plants with Seeds 35

Chapter 6: Angiosperms I: Flowers Aren't Just Pretty! 43

Chapter 7: Angiosperms II: The Amazing Diversity of Flowering Plants 51

Chapter 8: Non-plants and Strange Plants 59

Chapter 9: Plants as Food 67

Chapter 10: Plants Fight Back! 75

Chapter 11: Plant Communities and Ecosystems 83

Chapter 12: Helpful Plants 93

Conclusion 99

Amazing Plant Facts 100

Key Terms 106

PREFACE

When I think about the scientific study of the natural world, two phrases from the writing of Pope St. John Paul II come to mind:

(1) a rigorous pursuit of truth and

(2) a love of learning.

The first—a rigorous pursuit of truth—describes science and its processes. Scientists make careful observations, design experiments, and collect data to learn more about how the world works. Too often, though, science may seem like something you do in a big research facility with a lab coat.

But we are all scientists!

Anyone can study the living world in a scientific way. From an early age, everyone has a curiosity to understand the world. Think of a baby repeatedly dropping something onto the floor; they are discovering how gravity works! It is this basic curiosity that drives science.

The second piece—a love of learning—also describes what science should inspire. Sometimes science is depicted as a dry, boring set of facts, but nothing could be further from the truth. The world is a fascinating place. I have been interested in the natural world my whole life, and this interest led me to obtain a BS in zoology and a PhD in ecology, evolution, and behavior. I have used this education to teach biology classes every day for a living, and yet I am still constantly amazed by the wonders of our world.

There is always something new to learn in all the natural sciences, so much inspiring beauty and wonder all around us. Life consistently exceeds my capacity to imagine it.

For example:

- Did you know there is a tree in California that has a trunk that is over one hundred feet around?

- Did you know there are some mushrooms that are longer than blue whales?

- Did you know the horsetail plant, given its name because of its resemblance to a horse's tail, have spores that can walk?

How can we help but love learning about this fascinating world we inhabit?

Finally, it is too often assumed in our society today that faith and science act in opposition to one another, that somehow if we learn enough about the world, it would disprove the existence of God. But it is important for each of us to be confident in our Faith and the fact that truth cannot be in opposition with itself.

Armillaria ostoyae (below), also known as the honey mushroom, is considered the largest and oldest organism on Earth.

> *"[Science and faith] each can draw the other into a wider world, a world in which both can flourish."*
>
> —Pope St. John Paul II in *Physics, Philosophy and Theology*

We read in the *Catechism of the Catholic Church*: "Methodical research in all branches of knowledge, provided it is carried out in a truly scientific manner and does not override moral laws, can never conflict with the faith, because the things of the world and the things of faith derive from the same God. The humble and persevering investigator of the secrets of nature is being led, as it were, by the hand of God in spite of himself, for it is God, the conserver of all things, who made them what they are" (CCC 159).

Holy Mother Church teaches us that we can pursue scientific knowledge unafraid. It is my hope that *The Foundations of Science* series will not simply give your family some facts about the world but instead instill a curiosity and love of learning in you that you can apply across all the disciplines of your life, both scientific and otherwise.

Timothy Polnaszek, PhD

HOMALOMENA WALLISII

INTRODUCTION

As you can gather from the title, this book is all about plants! On the pages that follow, we will tour the diversity of the plant kingdom and explore the many ways biological communities depend on plants.

The book starts out by describing some key features that make a plant a plant, and then it describes different major groups of plants. Along the way, we will identify other features certain plants have that allow them to carry out specific functions which help them survive and reproduce (like seeds and flowers!). Throughout the book, but especially in the last chapters, we also point out the many contributions of plants to our world. A big "Thank you!" to all the plants out there for the oxygen we breathe and the food we eat!

Plants and their role in the natural world are expansive topics, though, and so we could not include every amazing aspect of plant-life in this book. But I do hope that this book generates discussions about the natural world and inspires further exploration.

Plants: Wonders of Nature is structured such that it can be used independently from the other units in *The Foundations of Science* series, but it does provide a good complement to the exploration of animal life found in *Animals: Creatures of the Wild*, and it introduces some ecological concepts that will be explored more fully in *Ecosystems and Species: The Web of Nature*.

The siam tulip, also called a summer tulip, is not actually related to the tulip, but to the various ginger species, such as turmeric.

CHAPTER 1

WHAT IS A PLANT, ANYWAY?

CONSIDERING WHAT MAKES A PLANT A PLANT

The group "plants" includes many organisms we see every day—things like trees, grass, and bushes. You probably know a plant when you see one, and it is usually pretty easy to find one out in nature. Even in the middle of a big city like New York or Chicago, you can still find plants.

But grass is pretty different than a tree, so what makes a plant a plant?

When I am teaching, I sometimes ask students to answer this question, and they often say things like "plants are green" or "plants have leaves," but as you may or may not know, there are some plants that are not green and some that don't have leaves.

Biologists have been organizing species into groups since the time of the Ancient Greeks (and likely even further back in time). Aristotle, a famous Greek philosopher, divided living things into two groups: animals and vegetables (this second one being organisms that do not move, like plants and fungi). One major feature he and others used to separate animals from plants is the ability to move. In the *Foundations* series animal textbook, we talk about movement

Plants are remarkably diverse. They include corn, shrubs, waterlilies, and oak trees.

(to chase dinner, run from predators, or find shelter) as a defining feature of the animal group. It is also true that plants do not move. So was Aristotle right?

Not entirely.

It turns out there are living things that do *not* move around that are *not* plants (like mushrooms). Just like any science, we add information over time to better understand the world around us, including adding information to how we define groups of living organisms. What other features should we add to define the plant group? Let's find out!

THREE FEATURES OF PLANTS

If a living thing is defined as a member of the plant group, it has a few key characteristics. One is that they are usually **multicellular**. Some organisms are made up of only one cell (**unicellular**) and others have many cells, including plants. All the plants we find on land are multicellular. This rule may not be perfect, as it would leave some single-celled algae out of the plant family even though they do share some plant-like characteristics (we'll check back in on these odd 'might-be-plants' in chapter 8).

Another trait found in plants is that their cells have walls made of a tough material called **cellulose**. This separates them from both animals and fungi; animals' cells have no cell walls, and fungal cells have a cell wall made up of a different material (**chitin**). Cell walls help support individual cells but also give the larger organism (the plant we see) some structure as well.

The last feature we will mention here is an important one: plants can conduct **photosynthesis**. We will talk more about photosynthesis in later chapters, but for now, just know that it means plants can generate their own food from sunlight. (Photo = light, and synthesis = to combine or make, so plants make food from light energy!) If you read our book on animals, you know this means plants are autotrophs, whereas animals, who cannot produce their own food but must to go out to find and capture it, are **heterotrophs**. Photosynthesis is hugely important for entire ecosystems because this is where energy enters into the ecosystem; without plants, animals would have nothing to eat and entire food chains would fall apart. **Food chain** is the term we give to the tiered series

Remember: Photosynthesis, photo = light, and synthesis = to combine or make, so plants make food from light energy!

of organisms that are all dependent on the next as a source of food (animals eat plants, we eat animals, etc.). Another amazing feature of photosynthesis is that plants release oxygen as a part of the process, and we breathe in the oxygen that plants "exhale." Without plants, the world would be a wildly different place.

PLANT DIVERSITY

Now that we know a little bit about plants, let's take a quick look at some examples and interesting facts to help us introduce the rest of this book.

There are around 375,000 species of plants described on earth, and likely even more we haven't even discovered yet! Major plant groups include mosses, ferns, conifers (pine trees), and many types of flowering plants. These plant types are quite diverse. Mosses, for example, only grow close to the ground and in moist environments (we'll see why in a later chapter). One type of flowering plant you may know, the saguaro cactus, looks quite different than moss (much taller, and it has spines for protection) and survives in harsh, dry conditions, like deserts. In this book, we will take a look at some of the **traits** (or distinguishing qualities, physical characteristics) that plants have that enable them to survive and thrive in different environments.

For now, though, let's just take a closer look at the extreme size differences possible between plant species. Duckweed, a pond-living plant, is one of the smallest plants in the world, at only a fraction of an inch in size. In contrast, the sequoia trees (or redwoods) can be more than three hundred feet tall! The most well-known tree in the world is named General Sherman, found in the state of California. Its trunk is 103 feet around and it is over 50,000 cubic feet in volume; that means, by very rough approximation, it would take a fleet of 400 to 500 minivans to haul it away if it fell in the forest, with all the seats folded down!

For their part, plants also come in many sizes between those two extremes, and both large and small are found basically everywhere on earth, with very few exceptions. We even see plants in the ocean and in Antarctica. The types of plants in each area will depend on the habitat found in the area. Things like how hot it is and how much rain falls are pretty important in determining how many and what types of plants there are in each place.

The Plants We Eat ... in One Way or Another

Make a list of everything you ate so far today. Can you find anything on that list that doesn't have plants as part of the ingredients? Even for those things that aren't made up of plants (like milk to drink), we still needed plants to get that food to the table. After all, a cow can only grow up to produce milk by eating plants!

Plant species are not only found out in nature, though; we also bring them into homes as house plants and rely on them as crops to feed us (and our animal friends). The world food supply relies on 150 or more plant species—without these plants turning light energy into their own food, we wouldn't have any food. About half of the energy we humans get from food comes from three major crops: rice, wheat, and corn. These and other crops have been grown by humans for a long time. The first cultivation of plants for food may have happened over 20,000 years ago! We didn't develop agriculture as we know it today, with large fields and farms, until much later, but still thousands of years ago.

Overall, plants are an exciting and important part of our earth's ecosystems and our own everyday lives. It's hard to imagine what the world would look like without plants. In the rest of these chapters, I hope you will find a fun and factual tour of these natural wonders that fill our world.

Some of the sequoias (redwoods) found in California are so massive they can make human beings look like insects!

Plants are found even in the desert, though it is more difficult for them to survive in such a dry climate.

FOUNDATIONS REVIEW

✓ There is a wide diversity of plants in the world, and not all are green or have leaves. We cannot even say that plants are living things that don't move, because there are some organisms in the world that do *not* move and are *not* plants.

✓ Therefore, we have three distinct features that we use to identify a plant: they are multicellular, meaning they have more than one cell, their cells have a tough outer wall of protection made up of a material called cellulose, and finally, plants have the ability to generate their own food using sunlight through a process called photosynthesis.

✓ Plants grow in places all over the world, including even in the ocean and Antarctica, and they come in many different sizes, from less than an inch to over three hundred feet tall. There are over 375,000 species of plants in the world, and we depend on them for our very life, not just because we eat many of them, but because they support entire ecosystems of life, including many of the animals that sustain us.

among other amazing plant facts and phenomena. They are also critically important to the biological communities around them and a big part of our lives, giving us food, fuel, medicines, and more.

Biblical authors, and Jesus himself, also saw plants all around in the natural world, using them for food and for building materials and so many other things. Because plants affect so much, both now and then, it's unsurprising that plants feature prominently in the Bible in many places. Consider the cedars of Lebanon, the branches of palm trees used on Palm Sunday, bitter herbs eaten at Passover, lilies of the field in the Sermon on the Mount, and the parable of the mustard seed, to name just a few.

nected than you might think, and it can be fun to think about some things that are unchanged, like our reliance on plants. Jesus, as God made flesh, also lived in our plant-filled world. I wonder if he had a favorite tree for shade growing up? Or worked with his parents to grow fruits or vegetables in a garden? What do you think?

Beams of sunlight fall between the canopy of a massive beech tree. The interaction between sunlight and plants is an integral part of supporting life on earth.

CHAPTER 2

PLANTS' SUPERPOWER
Generating Food from Light!

PANDO

In a forest in Utah, there is a large grove of more than 45,000 aspen trees. This may not be all that remarkable, until you realize that each tree is genetically identical. It is thought that each of the trees has grown outward as a clone from a single individual. The trees even have connections to each other underground. So rather than a single tree, it is more like many trees linked together to form one individual, or one "super-organism." This "individual" even has its own nickname: Pando. Together, these trees weigh more than 6,500 tons!

How did Pando grow to be so big? The short answer is: the sun! It is the sun that gave it the energy it needed to grow and produce all that biomass. But let's expand on that answer.

Pando grows in Utah, in Fishlake National Forest. It is also known as "the trembling giant."

PHOTOSYNTHESIS = ENERGY!

In the last chapter, we mentioned **photosynthesis**, the process by which plants make their own food. In this chapter, we will take a closer look at this important ability and how it actually works.

It is hard to exaggerate how important photosynthesis is to life on earth since this is how **energy** enters into the **food chain**. Energy is a tough word to define even though we use it all the time, and there are lots of types of energy, but to keep it simple, we can say that energy gives things (in our case here . . . *living* things) the ability to "work," or to function and stay alive, just as batteries might give a toy the "energy" it needs to work. If you think about your own life, you need physical energy to go out and play, and mental energy to do your school work. You get this energy mostly from food because food is where energy is stored. If you skip a meal, your energy level will drop, and you will become tired (and grumpy!). All organisms, including you, break down food molecules to generate the energy they need to live and thrive.

Well, you might ask, how does that energy get into our food? Photosynthesis, of course!

Through this process, plants generate their own food using the sun's energy. We of course eat some plant life, but we also eat the animals that eat the plants, so we get our energy that way too.

Photosynthesis is a great example of God's creativity highlighted in nature. Plants (and other organisms) harness the abundant energy provided by the sun, and in turn use it to live and grow, which allows us to live and grow, as well as all those wonderful creatures we read about in our *Animals* textbook. Since photosynthesis is so important to life on earth, it's worth a more in-depth look.

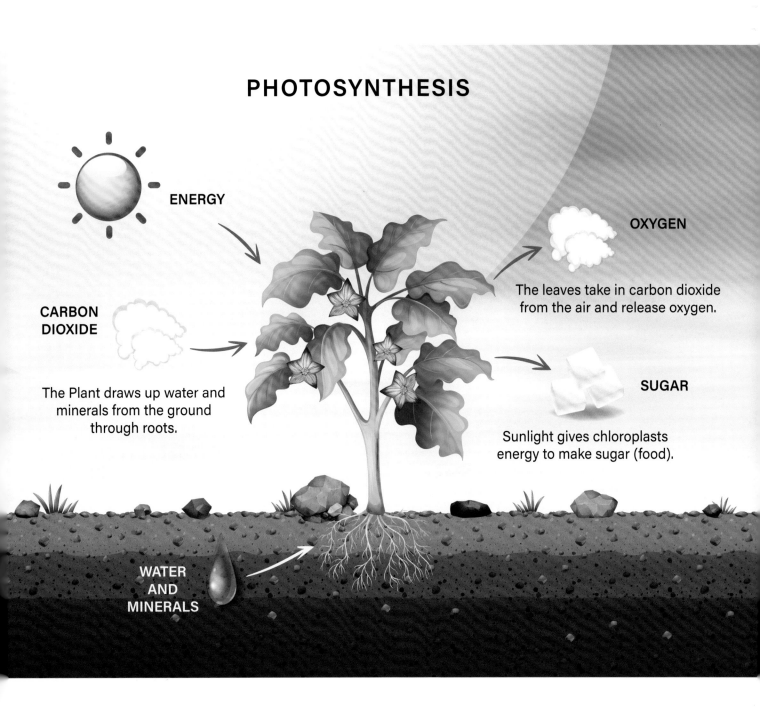

TURNING LIGHT INTO FOOD

How do plants capture light energy and turn it into food? The answer to that question is important to understanding how life on earth is able to survive. The first step would be to absorb the energy that is found within light. Whenever light hits a surface, some is reflected and some is absorbed. Asphalt in a parking lot heats up over the course of a day because the energy in light is being absorbed by the blacktop. Absorbing light can't be the whole story, though; after all, non-living things (like asphalt) and living things (like animals) absorb light, but they don't generate their own food from that light like plants can. (An animal cannot "eat" light to live, right?) The key is a special little structure that plants have in their cells called **chloroplasts**.

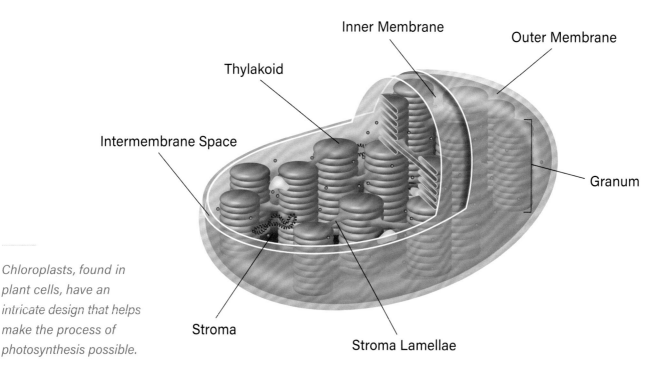

Chloroplasts, found in plant cells, have an intricate design that helps make the process of photosynthesis possible.

Plants and animals are both made up of many cells. Within these cells, we find many specialized compartments that do different jobs. For example, in your body's cells, there are things called lysosomes that function like little garbage collectors; they absorb waste, break it down, and can dump it outside of the cell. We can explore all these compartments (called organelles) and their jobs in more detail in another book. For now, let's look at chloroplasts, an organelle that plants have, but animals do not.

Inside the plant cells, these chloroplasts not only absorb light, but also capture the energy that light contains. This is made possible by small molecules called pigments. One of the most important pigments is a green one called **chlorophyll**. Chlorophyll is actually what gives plants their green color. When light hits these chloroplasts, they absorb most of the light but reflect the light, which our eyes detect as a green color. Some rare plants that do

Animal Thieves

Animal cells do not have chloroplasts, so they cannot directly turn light energy into food. But because biology is so much weirder than you can imagine, there are animals that actually steal chloroplasts from other organisms to use them for themselves. That's right, they steal energy-producing factories so they don't need to find and eat as much food themselves! The eastern emerald elysia starts eating algae early in its life. Rather than chewing up the algae, it sucks out the contents of the algal cells like sipping from a straw. Then, the chloroplasts are stored in the body of the sea slug where they can continue to collect sunlight and create sugar.

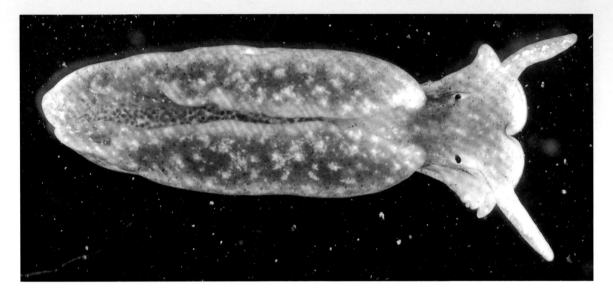

not conduct photosynthesis have very few or no chloroplasts, so guess what? They aren't so green!

But let's return to chlorophyll. Pigments like chlorophyll absorb light energy and then plants turn the energy into a more usable form called chemical energy. You might think of it like how your parents, in a sense, turn money into food. Money is great, but you can't eat it, right? So your parents have jobs which help them "capture" money, and they take that money and go purchase food to eat. In this analogy, plants use the chlorophyll to capture light energy (their "money") and then they turn that into chemical energy (food to eat).

Let's get away from that analogy though to discuss what chemical energy really is. As a quick explanation, everything is made up of atoms (a water molecule, for example, is made of two hydrogen atoms and one oxygen atom … H_2O), and whenever atoms are bonded together, it stores energy in those bonds. When these chemical bonds between atoms break apart, it releases the energy. In fact, we get energy from our food by breaking down the chemical bonds in things like sugars and proteins. So light energy is essentially harnessed and stored by plants in food molecules it can digest later.

OK, so far we have chlorophyll, found in chloroplasts, that help plants harness the energy found in light. What else do plants need to successfully make their own food? Here is an equation describing the process of photosynthesis—think of it like a recipe: everything on the left (ingredients) is turned into the things on the right (products), just like flour, sugar, and other ingredients produce a cake.

Light + water (H_2O) + carbon dioxide (CO_2) → Food (sugars) + oxygen (O_2)

So plants need light (captured by chlorophyll), but they also need water (H_2O) and carbon dioxide (CO_2) to create their own food. Let's look at how the plant gets each of these other key ingredients in their recipe.

HOW PLANTS GET WHAT THEY NEED

We already talked about how plants get light, so let's talk about water now. Water enters plants via their root system (if a plant has roots, that is, because not all do) through either rain or through our efforts to give them water. Absorbing water is critically important to plants because they can't conduct photosynthesis without it. If your family has ever forgotten to water a house plant, you will know firsthand that plants do not do well without water.

Plants also need carbon dioxide. There is a lot of carbon dioxide in the air surrounding us. We take in air (oxygen) by breathing, but do plants breathe? It turns out that plants have small openings called **stoma** (plural is **stomata**) to help them exchange gasses with the atmosphere. Look at the photosynthesis equation again:

LIGHT + WATER (H_2O) + CARBON DIOXIDE (CO_2) → FOOD (SUGARS) + OXYGEN (O_2)

Notice there are two gasses present in the equation. Carbon dioxide (CO_2) is on the left side of the equation, but oxygen (O_2) is on the right. This means that plants use up carbon dioxide during photosynthesis and then generate, or put out, oxygen. Two kidney bean shaped cells next to the stoma help the plant open and close this opening to regulate its "breathing in" of carbon dioxide and its "breathing out" of oxygen. This is why plants are so important to human beings, because we need oxygen to live.

Now that we know the basic recipe for creating food from light, how does it actually happen? To store the energy from light, the next step is to actually assemble food molecules, or sugars. We already mentioned that atoms make up the world around us, and the same is true for sugar molecules. Sugars are basically big strings of carbon, oxygen, and hydrogen. Carbon dioxide has carbon in it, and plants disassemble this molecule (like taking apart tinker toys or Legos) and reassemble the atoms, along with hydrogen from water, into a sugar molecule. This whole assembly process is fueled by the energy originally captured by chlorophyll, but now it is stored in food molecules. Once the food is made, it can be stored up by the plant to use later for growing, repairing wounds, producing seeds, or other important things plants need to do.

WHAT TO REMEMBER

Photosynthesis is a complicated process. And guess what? There are many more details to these chemical reactions and processes than what we covered here. But now you know the basics of how this amazing ability is carried out, one that almost all plants conduct, and it is one of the key features of this group of organisms.

Even if you struggle to remember all these difficult words, understanding the general process of photosynthesis—of how plants can use sunlight to make their own food, and that in turn benefits all life on earth—is the important thing to remember!

In our next chapters, we will start to explore different types of plants and the features that define each of the plant groups.

No plants? No oxygen!

What do we use from the air we breathe in? Oxygen! Where does all this oxygen in the air come from? Plants! When plants make their own food, surplus oxygen is released into the air.

In this oxygen molecular model, the atoms are represented by red spheres. Since Oxygen = O_2, there are two of them.

FOUNDATIONS REVIEW

✓ Photosynthesis is incredibly important to all life on earth because it helps sustain the food chain by injecting energy into it. Animals (including us) would not be able to receive the energy they need to live if plants did not capture energy from the sun.

✓ The key to photosynthesis is special little structures that plants have in their cells called chloroplasts, which help them absorb light energy. They do this through pigments like chlorophyll, which absorb light energy, and then plants turn the energy into a more usable form called chemical energy.

✓ Plants also need water to live, which they get through rain or our efforts to water them, in addition to carbon dioxide (CO_2). When they take in the carbon dioxide, they put out oxygen, which we need to breathe, so in this way plants are vital to our survival.

Photosynthesis and Grace

All organisms need energy to live and thrive. We just discussed how plants generate their own food by using solar energy—photosynthesis! This makes photosynthesis a fascinating and critical part of earth's ecosystems. Many organisms cannot make their own food, and so instead rely on eating other organisms to obtain energy. But wherever you are in the food chain—rabbit eating clover, fox eating rabbit, etc.—the energy that is being passed along needs to originate *somewhere*. For most ecosystems, that source is photosynthesis. There are also rarer organisms (some bacteria and others) that can use energy generated from chemical processes, and these form the base of food chains in places without sunlight (like the deepest depths of the oceans). Either way, only once that energy is initially captured can it be passed throughout the ecosystem.

There's an interesting analogy here to God's grace and how it works as it spreads throughout the Body of Christ. The *Catechism* says grace is how we become "partakers of the divine nature" (CCC 1996). This means grace is like God giving us a little bit of his

very life. In a sense, it's the "spiritual energy" our souls need, just as our bodies need physical energy.

Of course God gives each of us his grace chiefly through the sacraments, but his grace can come to us through other people as well, *through the grace he has given them*. For example, think about how much you have benefited from the grace God has given your parents throughout their lives. As they raise you and teach you about the world, leading you on a path to heaven, they are able to do so because God has given them his grace. This is similar to how energy enters the food chain through photosynthesis. Just as the sun is the source of the energy captured in photosynthesis, God is the source of all grace.

So let us never forget that as children of God instructed to love our neighbor, we must cooperate with God's grace and receive it into our souls, in order that we might pass on that "spiritual energy" to others. What can you do today to pass God's grace through the Body of Christ?

A waterfall cascades over rocks blanketed by moss, one of the most common types of bryophytes.

CHAPTER 3

BRYOPHYTES
Living the Simple Life

NO FLOWERS? NO SEEDS? NO ROOTS? NO PROBLEM!

The easiest way to discuss major plant groups is to look at the key features that separate them. We did the same thing with animals in the *Animals* textbook, organizing them by the features they share so that we had chapters on birds, fish, mammals, reptiles, and amphibians. Some of these defining features of plants include whether or not the plant has flowers, seeds, vasculature (which is a transport system we'll discuss in more detail), and specialized parts like roots for taking up water and leaves for conducting photosynthesis.

The most basic plant group, and the topic of this chapter, lacks all of these features. They are known as **bryophytes**.

So what does a flowerless, seedless plant without true roots or leaves actually look like? The most common examples from this group are mosses, which you may be able to find growing near your own home, but we'll also meet some other lesser known members of this group. These plants generally grow in poor soil and near water or in moist, shady areas. Before we learn more about plants in this group, let's use them as an example to think about some of the challenges of being a plant.

CHALLENGES TO SURVIVAL LIVING ON LAND

Living on land comes with specific challenges that do not exist for those plants living in water. As we mentioned in the previous chapter, plants need water to go through photosynthesis. Plants living in water have it available to them all the time. But what about plants living on land? Life on land means you can lose too much water and dry out. To counter this, some plants have waxy substances or tough outer layers to prevent water loss, sort of like how raincoats or other waterproofed materials work, but to keep water *in* instead of *out*.

Plants that grow in drier areas have distinctly different challenges to survival than those that grow in wet and moist areas.

Another challenge is that the plant needs to "breathe" differently. Just like fish have gills to get oxygen from water and mammals have lungs to breathe air, a plant would need different features in each place to get the CO_2 it needs for photosynthesis. We already learned about stomata that help with gas exchange, and so you probably won't be surprised to learn that almost all bryophytes, and other land plants, have stomata to help them "breathe."

Another challenge to living on land, but an important one, is gravity. Why would gravity affect a plant? Well, have you ever considered that water is heavy? A gallon jug of water weighs almost 8.5 pounds. We already mentioned that land plants need to avoid drying out, but to do this, they need some way to move water up from the ground to all the parts of their body (plants in water would not have this issue because, of course, they are surrounded by an abundance of water).

One way that other plants move water is with **vasculature**, or a transport system, which is like a series of tubes, similar to our own blood vessels that move blood throughout our bodies. But bryophytes do not have vasculature; they are called **non-vascular plants**. If we know that they do not have a specialized system to move things very far (like water), how tall do you think a bryophyte can grow? Would it be more similar in height to grass, a bush, or a tree? We will come back to this question when we look at some bryophyte examples. For now, though, let's step back from bryophytes temporarily to look at similar plant characteristics more broadly.

Plant Fun Fact: Plants grow the most at their tips, both top and bottom.

SIMILAR CHARACTERISTICS IN PLANTS

At this point you may wonder why bryophytes are plants when they don't have flowers, seeds, roots, and true leaves. What makes us so sure that they are plants? After all, a tiny moss spikelet looks dramatically different than a giant oak tree, or even from a rose bush. What features do they share? We already mentioned some in the first chapter (like photosynthesis and multicellularity), but what else is the same between groups of plants?

One thing land plants have in common is **apical meristems.** That just means that plants grow the most at their tips, both top and bottom. Multicellular organisms grow larger by having *more cells*, not by having *their cells grow larger*. As we humans grow up, we sort of grow in all directions at once. Instead, a lot of a plant's growth is concentrated at the top and bottom; for example, roots continuously spread out from the tip and branches also grow outward and upward. Cells divide more rapidly at these meristems, and that creates growth—more cells result in a bigger size (like longer roots or a taller tree).

The next thing we will mention here is that plants have **multicellular, dependent embryos**. This is simply saying that the new, tiny plants of the next

ORANGE DAYLILY

generation get their start while still with their parent plant (sort of like how you started your life in your mother's womb). Some animals may lay eggs and abandon them, but plant parents help out the next generation. We can see this in many food plants that grow in our gardens. Seeds in tomatoes, cucumbers, and other plants spend time with their parent plant (inside the fruit in these examples) and then eventually move away from their parents and grow on their own.

One last thing plants have (that animals do not) is something called **alternation of generations**. Plants often have two "forms" (or types) that each reproduce to develop into the other form. One way to understand this is if butterflies were to lay eggs which developed into caterpillars, and then (rather than directly turning into butterflies by changing in a chrysalis) caterpillars laid eggs which developed into butterflies—two separate types that each develop into each other. Ferns may be the best example, since the two types live independently; a fern frond is one form, and a small heart-shaped plant is the other. In bryophytes, one form usually develops and grows right on top of the other (so it is dependent on its parent plant). Bryophytes may be odd plants that lack certain features, but they do have two alternating generations, they grow at their ends with meristems, and they rely on parent plants, like all other plants do. So yes, they are plants!

Ferns are one of the best examples of the phenomenon among plants of alternation of generations.

TYPES OF BRYOPHYTES

Now that we have discussed both similarities and differences with other plants, let's take a quick look at some of the types of non-vascular plants.

Mosses

First, we will take a look at the "true" bryophyte group, the mosses (~12,000 species). Moss grows in moist, shady environments as a mat of small, spiky plants. When looking at mosses along the forest floor, you may also notice taller stalks growing with a pod or capsule at the end. The spikelet and the stalk are the two forms, or generations, of the moss. The capsule holds spores, which is what mosses and other bryophytes use to reproduce instead of seeds.

Let's return to our question about bryophyte size from a moment ago. Because they lack an efficient way to move water around, bryophytes cannot grow very tall. (Is that what you predicted?) In fact, the tallest of any bryophyte is a moss called *Dawsonia superba,* which is less than two feet tall. Most other mosses only grow less than two inches in height.

Hornworts (or Anthocerophyta)

Next, we will look at the hornworts, or anthocerophyta. There are about three hundred species worldwide, and all of them lack flowers and seeds, like other bryophytes. They resemble small, flat, round, leaf-like plants that grow along the ground. Hornworts are named for a horn-like spike that grows straight upwards away from the small round plant. This horn contains spores, which will grow into the next generation of hornwort plants. As the horn dries out, it splits open and releases the spores away from the parent plant. Each spore then develops into a new plant. Hornworts lack true roots, but, like other bryophytes, they have small hair-like structures called **rhizoids** to help themselves anchor onto the ground.

Liverworts (or Hepatophyta)

The last group, liverworts, or hepatophyta, are represented by around nine thousand species worldwide. They are small plants that grow along the ground in one of two forms: either flat, leaf-like ribbons (more like hornworts) or a branching leafy shape (more like mosses). Liverworts grow little umbrella-like structures that will develop and release spores, serving

Still Trying to Figure It Out

At times we say a plant is a "true" bryophyte because scientists are still testing several ideas about how these groups of plants are related to each other. Some evidence suggests they should all be in one group together, but other evidence suggests we should break them up into smaller groups (so separate hornworts from moss and liverworts, for example). Scientists will continue to look at similarities in appearances, DNA, and many other features to try to determine the correct grouping of these plants. This is a great example of how science is always looking to learn more and update our ideas about the way the world works.

THALLOSE LIVERWORT

the same purpose as the "horn" on a hornwort. The plants in this chapter generally rely on moist habitats and grow near water, but liverworts are the *most* dependent on water (they have more trouble with drying out and need water to send out their spores). Some liverwort species can regrow from a small piece—you could break a liverwort into two pieces and end up with two plants!

With our discussion of bryophytes complete, we'll turn in the next chapter to ferns and all their friends.

FOUNDATIONS REVIEW

- ✓ Bryophytes are a kind of plant that has no seeds, flowers, or true leaves or roots. These plants generally grow in poor soil and near water or in moist, shady areas. These include mosses, liverworts, and hornworts.

- ✓ Bryophytes do not have vasculature, which is like a transport system for moving water. This means they tend to be shorter because they cannot move water up against gravity very well.

- ✓ Because they lack common features that we usually think of when we think of plants (seeds, flowers, roots, etc.), they may not seem like plants. But they are because they share other traits with plants, including having two alternating generations, growing at their ends with meristems, and relying on parent plants.

The Parable of the Sower

This is a book about plants, so eventually we had to mention Jesus's parable of the sower and the seeds found in the Gospels. You may know this story, but as a reminder, a sower spreads seeds onto the ground in various places: onto a path, onto rocky soil, among weeds and thorns, and on good soil. The success in growing depended on where the seeds landed.

Here, the symbolism of the seed is God's Word, and we are the soil. We have the opportunity to let his Word grow in our hearts and bear good fruit through our actions. Seeds that landed on the path were carried away by birds—if we aren't receptive to God's Word, the message cannot take root. If seeds land on rocky soil, the roots do not grow deep enough. This means that we may welcome the teachings of Jesus into our hearts, but at the first sign of hardship or suffering, we abandon our faith very easily. If seeds try to grow among the thorns, they are choked out by the plants growing around them. This is like letting the world drown out God's Word. Only the seeds that fell on good soil grew to bear fruit.

Here in this parable, Jesus uses analogies that we can actually observe in nature. When seeds (or the spores of bryophytes) disperse or are planted by a gardener, the soil and the animals and plants around these young plants can affect how they grow. We just learned how bryophytes do not have many advantages compared to other plants (no vascular system, no roots or leaves, etc.), so how do they prevent being choked out by some of these other plants? One clever way is to grow where other plants cannot— some bryophytes grow on fallen logs, rocks, and in poor soil, even sometimes on top of animal dung (dung mosses).

The fiddlehead of a young fern, a common vascular plant. Vasculature helps move water and other nutrients throughout the plant.

CHAPTER 4

FERNS AND FRIENDS
Vascular, Seedless Plants

LEVELING UP!

The plants in this chapter have "leveled up," and are now complete with fancy vascular systems. What do we mean by leveling up? Well, think of it this way. In a video game or board game, if you are controlling a character, you might receive or be rewarded with new abilities or powers throughout the game that help you win. In the same way, the plants we will begin to look at here have the extra advantage of a vascular system that the bryophytes from the last chapter did not have.

Remember that moss grows low to the ground because they don't have an efficient system to transport water (and nutrients and other things) around the plant. But **vascular plants**, meaning plants that have this tube-like transport system, also have other features we typically associate with plants, like roots, stems, and leaves. These are specialized regions of the plant that do certain jobs and are made up of tissues (similar cell types working together) or organs (collections of tissues that serve a particular function). For example, roots are an organ that contain cells and tissues that will help with taking in water from the soil, but leaf tissue wouldn't have cells with those same features.

As you can imagine, having traits like leaves, roots, and a vascular system provides vascular plants with distinct advantages. It may not be a surprise, then, that the most plant species on earth are vascular plants; eight times as many land plants are vascular, in fact, with over 300,000 vascular species, and more being discovered all the time. All the remaining plant groups we will discuss have specialized tissues and vasculature. Before meeting our example species from the vascular, seedless plants, let's take a closer look at these specialized regions of the plant and the vascular system.

A Plant's "Circulatory System"

An adaptation in biology is any trait that provides some advantage or plays a key functional role—perhaps making the organism well-suited to the environment it lives in or providing protection against predators it encounters. We see many marvelous adaptations all throughout this natural world God has given us. The vascular system present in some plants is an adaptation because it efficiently circulates water and other things within the plant, sort of like you have a circulatory system in your body that moves blood around through veins and arteries.

ROOTS

We will start at the bottom of the plant by looking at the **roots**. What sorts of advantages do roots provide to a plant? One is that roots can help anchor a plant into the ground. This is especially important if the plant grows tall, because then wind can become more of an issue.

But anchoring a plant is not the only job roots have. Plants need lots of minerals and nutrients in order to grow and stay healthy; things like nitrogen, phosphorus, and potassium. This is similar to how human beings need various minerals and nutrients in our diet. (Check out the nutrition labels on the food you eat; we need potassium, calcium, and many others.) Plants get these from the soil surrounding them, and roots help bring these nutrients into the plant's body. Sometimes this is simply by absorbing them, and other times the plant's cells play a more active role in bringing in these nutrients.

Roots help a plant soak up water and other nutrients from the soil, as well as anchor it. They usually remain underground, but some that grow above the earth allow us to see how elaborate root systems can be.

As we alluded to earlier, roots also absorb water from the ground to meet the plant's water needs. Since a plant can't move to find water and nutrients the way animals can, it needs to rely on its roots to grow through the soil to collect these things. In the Kalahari Desert, there is a plant called the Shepherd plant whose roots can go over two hundred feet deep into the ground!

Finally, not all roots look the same. Some plants have a central root that may help with anchoring and storage of nutrients; this is called a **taproot**. The orange part of the carrot we eat is a taproot of the carrot plant. Other roots branch out into a more tangled web of roots like what we are used to seeing; these are called **fibrous roots**.

LEAVES

Next, we'll journey higher on the plant by talking about leaves. One of the most important functions of the leaf is to conduct photosynthesis. Leaves are often packed with chloroplasts, those little sugar-making factories that are powered by the sun. In order to run photosynthesis, leaves also help take up carbon dioxide through stomata. But this brings about a dilemma, because opening stomata can also let out water vapor, which they also need for photosynthesis. Bigger, wider leaves can help capture more light, but may also lead to more water loss. Balancing all of these jobs has led to some interesting differences in leaf shape. Take the leaf of an oak tree and a needle from a pine tree; looking at them, it's difficult to believe they

OAK LEAF   PINE NEEDLES

Why do some leaves change colors in the fall?

We can also compare oak leaves and pine needles by their ability to withstand the cold. It turns out that a slim shape is better for extreme cold. This is one reason why **deciduous trees** (trees that drop their leaves and grow new ones each year), like oaks, let go of their leaves each fall and you get to rake them up with your mom and dad.

Another reason is that leaves are sunlight-capturing, energy-making factories. But in far northern or far southern areas, there is less sunlight for part of the year (approximately November to February in the north). If the "factories" are not productive anyway, it might make more sense to shut down production for the season.

But what color are the leaves when they fall to the ground? Usually not green, right? They change colors during the autumn season. We learned that chlorophyll is the pigment that catches light energy and also gives plants a green color. This means trees that lose their leaves each fall break down chlorophyll and store it elsewhere, because they don't want to dump the valuable chlorophyll onto the ground. There are other leftover pigments in the leaf that then give them their red, orange, gold, and brown colors. This gives us the glorious autumn colors we see each autumn!

are both leaves. The oak leaf is larger, so it can absorb more sunlight, leading to more sugar. But it also loses water faster than the slim pine needle.

Once the leaf has made sugars through photosynthesis, it needs to pass these food molecules to the rest of the plant. After all, what good is it if you can make your own food but can't send it to the rest of your body? Sometimes these sugars may be passed all the way to the roots for storage. This is where the vascular system comes in. This transport system is made up of several types of specialized, tube-like cells called **xylem** and **phloem** (pronounced zy-lem, like xylophone, and flow-em). Phloem helps move things from the top of the plant down, and xylem helps bring water and nutrients up. (Phloem flows down, xylem zips up!) This vascular system passes through the roots, stems, and leaves of the plant, allowing transport everywhere within the plant. For example, maple trees store things in their roots in the winter (sugar, water, nutrients) and sends those things up into the top of the tree through the vascular system; this provides everything needed to make leaves in the spring. We take advantage of this by harvesting some of the sap they move upwards to make maple syrup!

Again, most plant species on earth have these vascular systems for transporting things, including the ferns and friends that we will now look at in closer detail. But the plants in this chapter do not have seeds or flowers (we'll have more on those later). The major groups of seedless (and flowerless) vascular plants include ferns, club moss (which isn't a moss!), and horsetails.

TYPES OF VASCULAR, SEEDLESS PLANTS

Club mosses are found in a group called lycopods, which contains about one thousand species. These plants have small leaves, often small and pointy in shape, which make them resemble mosses. Because of this similarity, club mosses were given their name, even though we know now that they are different from mosses because they share adaptations with other vascular plants. Most existing species grow fairly low to the ground and are found in moist habitats. But some fossil species of lycopods were as large as trees! Fossils of club moss and other plants in this chapter are often found in coal deposits underground. Coal itself is actually formed from buried plant material by high heat and pressure. Any modern use of coal depends on plants, but ones from the distant past. Even when we don't realize it, we are depending on plants to help sustain our lives.

Horsetails sort of resemble a horse's tail (hence the name), but you may need to squint a bit or to give scientists a bit of artistic license in the naming here. They have a central stem,

Spiney Plants!

Not all leaves conduct photosynthesis. Cactuses have spines, which are essentially modified leaves that help protect the cactus from hungry animals. Instead, the whole trunk of a saguaro cactus has chloroplasts within it, taking over that role from the leaf (you can tell by the cactus's green color).

and then reduced, wispy leaves branching out all along the central stem. These plants are sometimes called "living fossils" because the ones we see today very closely resemble horsetail species found alongside ancient dinosaur fossils. There are only a handful of species found worldwide (~20). The most interesting thing about horsetails is that their spores can walk! The spores have four little spindly "legs" that curl and uncurl with changes in moisture, and the movement of the legs helps the spores move around (and even sometimes jump!).

Ferns grow mostly in moist, shady places around the world (though some can survive in dry environments), and there are around ten thousand species. Since they have more developed root systems and larger leaves, these plants are more similar to other plants than club mosses or horsetails. They are easily recognizable by their leaves, which are large, branching, and feathery looking fronds. As with other leaves, these hold the chloroplasts and conduct photosynthesis. As you may have guessed, these plants can grow much larger than bryophytes; in fact, the largest fern species can grow over thirty feet tall! Instead of seeds, they reproduce by **spores**. These spores develop in little yellow structures (called **sori**) on the underside of the leaf. Next time you find a fern in the wild, check out if it has any of these yellow dots under the leaf.

Overall, there are some really neat species of seedless, vascular plants, even though there are relatively few of these species on earth. Most of the species of plants that you'll find on a walk through the woods would be from the plant groups we'll take a look at in later chapters.

Plant Fun Fact: Horsetail plants are sometimes called "living fossils" because the ones we see today very closely resemble horsetail species found alongside ancient dinosaur fossils.

FOUNDATIONS REVIEW

✓ Unlike bryophytes, vascular plants have roots, stems, and leaves, as well as a vascular transport system that moves nutrients and water throughout the plant. This means they will be able to grow taller than bryophytes.

✓ Vascular plants have specialized regions that do certain jobs and are made up of tissues (similar cell types working together) or organs (collections of tissues that serve a particular function). For example, roots help anchor the plant and help it soak up water from the soil, while leaves allow the plant to conduct photosynthesis.

✓ There are several different kinds of vascular plants, with ferns being the most recognized and well-known. But there are also club mosses and horsetails, which receive their name because they resemble the tails of horses.

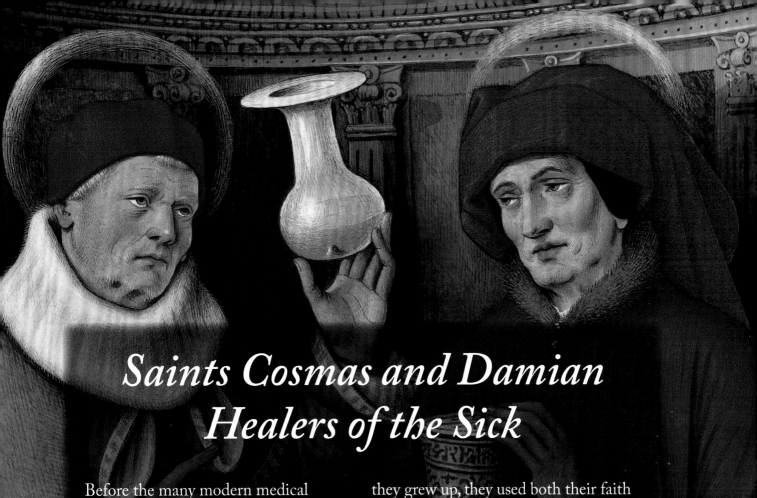

Saints Cosmas and Damian
Healers of the Sick

Before the many modern medical advances and technologies that we have today, plants were used as a primary source of medicine. Many plants contain compounds or chemicals that may have a healing effect, and historical doctors would use plants to develop medicines.

One example of this is the horsetails from this chapter, which were thought to help stop bleeding and treat wounds. Some of these traditional remedies have been confirmed by modern science, whereas others are less clear. Nevertheless, doctors used plants centuries ago, including two names you may recognize: Saints Cosmas and Damian.

Not very much is known about their lives, but Cosmas and Damian were twin brothers who likely grew up with a love for science and medicine. Their mother, Theodota, was also a Christian, and so raised them in her faith. When they grew up, they used both their faith and their reason to help them do great things for people. They traveled around the regions of Asia Minor (between the Black Sea and Mediterranean Sea) as healers, utilizing their knowledge of plants, minerals, and the natural world to make medicines and heal the sick. But they did not accept payment for their work, instead working as "unmercenary healers," which just means that they worked for the good of the person. The brothers also included prayer and teaching about God as a part of their ministry to the sick, knowing that it is important to treat the body but also the soul of a person. Because Christians were persecuted during their lifetime, we do know that they were eventually martyred for their faith.

Saints Cosmas and Damian, pray for us!

Pomegranates are a beautiful, red fruit filled with seeds. The term "granate" is derived from the Medieval Latin "granatum," meaning "many-seeded" or "containing grains." Each seed is encased in a sweet and juicy covering known as an aril.

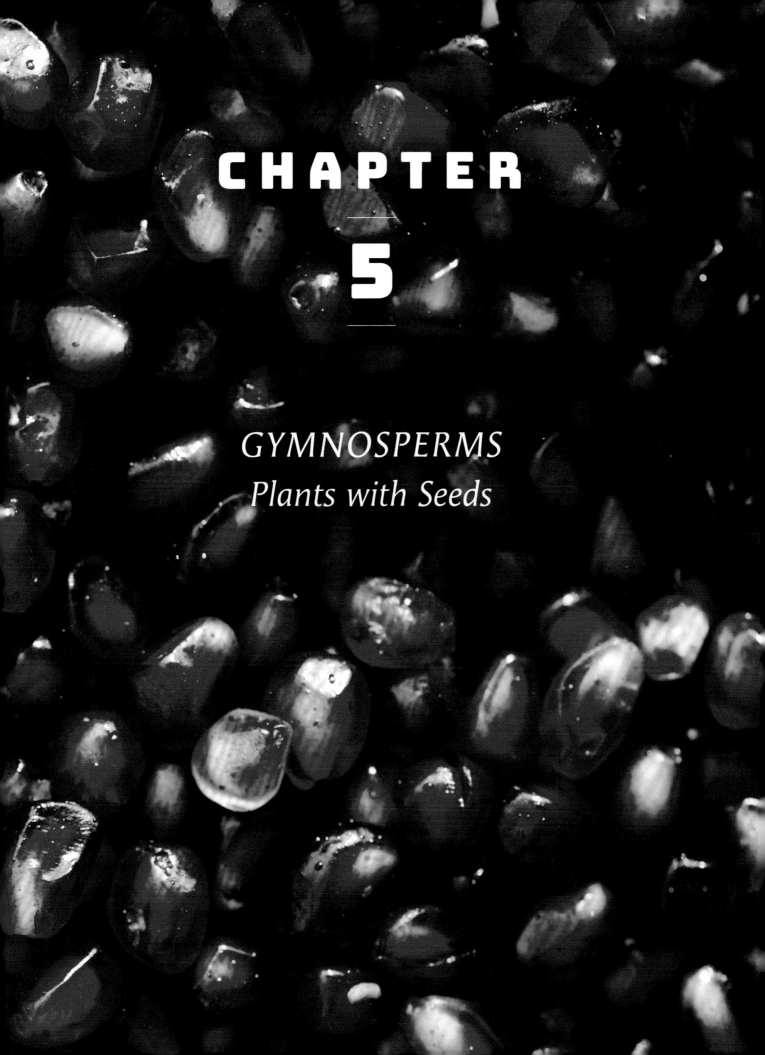

CHAPTER 5

GYMNOSPERMS
Plants with Seeds

LIFE WITH SEEDS

This chapter introduces yet another important feature in plants: seeds!

Gymnosperms, which include pine trees, share the trait of vasculature with ferns and their allies, but having a seed instead of a spore makes gymnosperm plants a little different. Here we have another example of how key features can help us categorize plants into groups. These features provide advantages to plants in certain conditions. We'll learn more about the features and advantages of seed plants in this chapter, but one advantage a pine tree (with seeds) has over a fern (without seeds) is that the offspring of the pine tree (the seeds) can survive in drier conditions. Seeds can survive drying out much more easily than spores. Most of the plants in the past chapters still depend on living near water, or in shady and moist conditions. Not so for gymnosperms, which consist of about one thousand species around the world, including some that can even live in dry deserts.

Before meeting some of the members of this group of plants, let's take a closer look at seeds and the characteristics that allows them to live all over the world in various habitats.

FACTS ABOUT SEEDS

What is a **seed**? Well, seeds are essentially a baby plant: they contain the plant embryo that represents the next generation as it grows into an adult. But seeds are also more than that.

Surrounding the seed is a **seed coat**. Just like you might put on a coat to protect you from the elements outside, a seed coat does a similar job for the seed. Though unlike one of our rain coats, one of its jobs is to keep some moisture *inside* the seed so that seeds can survive in dry conditions. The parent plant also packs the young plant a lunch, providing it with nutrients to thrive once it's ready to grow. In some plants (angiosperms, found in the next chapter), this is more prominent in a layer called the **endosperm**. The endosperm has nutritious food found inside the seed coat and outside the embryo, and the young plant can absorb these nutrients as it starts to grow. This process of a seed sprouting and starting to grow is called **germination**.

Once germination starts, the cells of small embryo inside the seed start to duplicate, which makes the developing plant grow larger and larger. At some point, it becomes too large for the seed coat and it bursts out. Two main parts emerge from the seed: one grows downward to become the roots, and the other grows up to become the stem and leaves.

You might wonder how a plant can know which way is up. After all, leaves would do no good buried underground where they can't absorb light. Believe it or not, scientists think that plants have cells that work as gravity detectors! Imagine a jar filled with water and rocks (or make one!). If you turn the jar over, the rocks will sink down to the new "bottom" of the jar. These plant cells are similar in that they have small "stones" that always sink to the

Plant Fun Fact:
Scientists think that plants have cells that work as gravity detectors, helping them to grow up.

ANATOMY OF A BEAN SEED

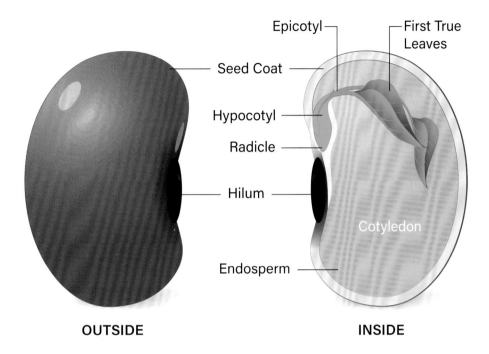

OUTSIDE   INSIDE

The structure of a seed allows for the baby plant to have the best possible chance to grow and thrive in the environment it's planted in.

bottom, telling the plant where to send its leaves, away from the "stones." Amazing, right?

Speaking of growing, seeds don't always sprout right away after they fall off of the parent plant. Unlike most animals, they don't have a specific amount of time they need to develop. Instead, the baby plant can sit, safe and snug, inside the seed coat until the conditions outside give it the best chance to grow. Some seeds need to go through a long period of cold before they germinate. By doing this, the seed won't sprout in the fall and so it can avoid freezing in the winter. Some species of plants (like lodgepole pine, a gymnosperm) can germinate after being exposed to high heat from a wildfire. Why would a plant want to grow after a fire? Well, if other plants have burned up, there will be less competition for water and nutrients.

On a similar note, since other plants compete with young plants for water, nutrients, and even light, it also makes sense for seeds to move away from their parent plant. Both the adult plant and the seed will have more success if the seeds can find a new place to grow. Being sent away can also allow the plant to expand where it lives into new areas and potentially find new habitats.

So how can seeds move? Some hitch a ride on animals—you may have encountered burrs that get stuck to your pants or a pet's fur. These are seeds that rely on mammals to get them to someplace new. Other plants rely on animals, but by trading something for the ride. Certain fruits that contain seeds are eaten by animals (free lunch for the animal!) but the seeds are then passed out the other end of the animal. Animal feces also makes good fertilizer (lots of nitrogen), so that's just an added bonus for the young plant. Another way to move around is by floating in the wind. Dandelion seeds have little parachutes to catch the wind and spread out the seeds.

Dandelion seeds can often be found floating in the wind, or in the gusts of "wind" blown from the mouths of little children!

TYPES OF GYMNOSPERMS

Now that we have taken a quick look at seeds and germination, let's look at the first of two plant groups that produce seeds: the gymnosperms. We'll take a brief look at three major groups of gymnosperms: conifers, cycads, and ginkgos.

Conifers

Conifers are the largest group of gymnosperms, with over six hundred species. This group contains those trees that we would call "pine trees," and also include redwoods, hemlocks, and cedar trees. The seeds of these plants are contained on cone-like structures, like pine cones you may have encountered in the woods before. The name *conifer* comes from Latin, and actually translates as "one that bears cones." Most of these plants are trees, and most have leaves year-round (also called "evergreen trees" because their leaves don't change color in the autumn or fall off in the winter). Many of the trees in this group have lots of sap or sticky resin, which also gives them a distinctive aroma. For the plant, this sticky sap helps protect them from insects that want to eat it and fungus that can cause infections. Like many gymnosperms, they are hearty plants that can survive in cold or dry climates.

Gymnosperms can be quite diverse, from conifers like this pine tree (left), to the cycad tree (right) found in more tropical habitats.

Cycads

Cycads resemble palm trees, but are gymnosperms (palm trees are angiosperms). These plants typically grow as one trunk without any major branches, and have a set of leaves near the top of the plant. The leaves also resemble palm leaves, with a central leaf stalk and branching fronds. Cycads are unlike many plants in that there are separate male and female plants. Most plants have both male and female structures; for example, pine trees have male and female cone types on the same tree, and flowers in the next chapter show both male and female parts. Cycad trees bear cones, like pines, but some plants only grow female cones (which will contain the seeds) and some plants only grow male cones (which contain pollen). There are only around three hundred species of cycads in the world, and several have recently gone extinct or are facing extinction.

Ginkgos

Ginkgo biloba is the only living member of the plant division called Ginkgophyta. In this way, it is sort of a living plant oddity. Most living species have closely related living relatives, but the ginkgo tree's only relatives are extinct and known from fossils. The living ginkgo tree closely resembles fossilized versions from the age of the dinosaurs, so I like to call ginkgo leaves living dinosaur food. The leaf has a rather unique structure that makes the tree easy to identify. All of the veins in each leaf run outward into a fan shape from the central stem rather than in a branching pattern. This plant has

Ancient Trees

Many conifers can live for a long time. A bristlecone pine tree, nicknamed Methuselah, is almost five thousand years old! It is recognized as the non-clonal tree with the greatest confirmed age in the world.

been used in gardens and landscaping for at least one thousand years, starting in China and spreading all over the world.

The ginkgo tree and the other gymnosperms all have seeds, which helps protect their offspring and disperse it away from the parent plant. But none of them have flowers, which is the trait we'll discuss in the next chapter.

Gingko tree leaves (right) have a distinctive and unique fan shape.

FOUNDATIONS REVIEW

- ✓ Gymnosperms share the trait of vasculature with ferns and their allies, but having a seed instead of a spore makes gymnosperm plants a little different. One advantage to having seeds instead of spores is that these plants can survive in drier climates, because a seed can survive drier conditions than a spore can.

- ✓ Seeds are essentially a baby plant: they contain the plant embryo that represents the next generation as it grows into an adult. It is surrounded by a seed coat, which protects it, and within this coat and outside the embryo there is a layer called the endosperm, which contains nutritious food so that the young plant can absorb these nutrients as it starts to grow. The process of a seed sprouting and starting to grow is called germination.

- ✓ There are several types of gymnosperms, including conifers (what we would call pine trees and also redwoods, hemlocks, and cedars), cycads (which resemble palm trees and are unique because they have separate male and female plants), and ginkgos (the *Ginkgo biloba* is the only living member of the plant division called Ginkgophyta).

A Dominican Scientist

In this chapter, we learned about germination and the early steps in the development of a young plant. Over the years, countless scientists have helped us better understand how plants develop and grow and about all of the tissues and organs that are vital to their ability to survive.

One scientist who helped with this process was a Catholic woman named Annie Chambers Ketchum. She was born in Kentucky in 1824 and was a school teacher, botanist, and scientist. In the 1860s, she moved to Europe and converted to Catholicism. It was there that she entered the Dominican order as a novitiate with the name Sister Amabilis. Later, she returned to the United States, where she continued her intellectual work as a scientist and Third Order Dominican.

Sister Amabilis was a botanist, but she also lectured on many topics, including science, art, and literature. In 1889, she authored a textbook called *Botany for Academies and Colleges: Consisting of Plant Development and Structure from Seaweed to Clematis*. It was a tremendous summary of scientific knowledge, containing more information about plants and their development and growth than other textbooks available at the time.

Sister Amabilis died in 1904 and was buried in her Dominican habit. She was a great example, alongside many other Catholic scientists, of someone who pursues greater knowledge in our faith and in the sciences, inspiring us to do the same.

The passion flower blossom is often used to symbolize events in the last hours of the life of Christ, the Passion of Christ, which accounts for the name of the group.

CHAPTER 6

ANGIOSPERMS I
Flowers Aren't Just Pretty!

EXTRA COVERAGE

The next two chapters cover the diversity of plants in the **angiosperm** group, also called flowering plants. These plants, as you likely guessed, have flowers! This characteristic is unique compared to the other groups we have looked at so far. Flowering plants represent more than 80 percent of known and described plant species, so it seems only fair to give them extra coverage.

In this chapter, we will focus mainly on the structure and function of flowers, as well as the advantages they provide. Then, in chapter 7, we will look at some of the many diverse plants found within the angiosperm group and examine some of their other non-flower features.

STRUCTURE OF A FLOWER

First, let's take a quick look at the structure of a flower, since some of these parts will become important when we talk about their function.

The outermost portions of a flower are the **sepals**. These are found just outside the **petals**, and typically look like small green leaves (though there is some variation). Before the flower is fully developed, these sepals can provide protection from cold weather or other things that could injure the flower.

The petals are the colorful part that make the flower pretty to look at; they are the reason we like to grow flowering plants in our gardens or use them for decorations. Petals can come in all sorts of shapes and sizes, depending on the plant. Inside the petals, we find the **stamen** and **pistil**. Unlike human beings, who are either male or female, these two parts of the flower are the male and female parts, both found within the same flower. The stamen is the male part of the flower and is made up of a **filament** (like a stem) and an **anther** at the top. The pistil is the female part and is often shaped like a vase. The top of the pistil is the **stigma**, the neck of the "vase" is called the **style** and the base is the **ovary** of the flower.

I know that was a lot of terms to remember, but now you know more about what makes a flower a flower!

Remember:
The sepal provides protection from cold weather and other things that could harm the flower.

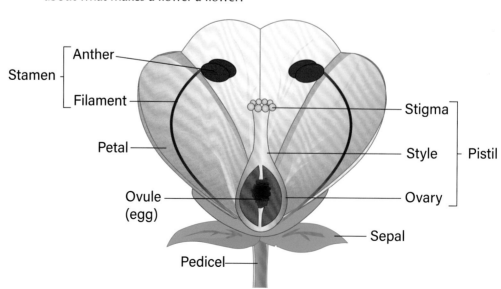

Bees and other pollinators engage in mutualisms with plants, where both organisms benefit, like a kind of trade.

THE ADVANTAGES OF HAVING FLOWERS: POLLINATION!

Flowers make it possible for pollination to take place. **Pollination**, which is the process of fertilization, or reproduction, in gymnosperm and angiosperm plants. It is called pollination because a pollen grain must be transferred from one plant to another for reproduction to take place and for seeds to develop. In gymnosperms, pollen grains are found on male cones and are transferred to the female cones (pinecones). But how do they get there? Mostly by floating on the wind. But this is a pretty inefficient process; plants that rely on wind dispersal for pollen sometimes must produce millions of pollen grains each year to make sure at least some reach the pinecones. Most of the grains simply blow away and eventually land on the ground (or make those of us with seasonal allergies sneeze a lot!). So wind is not always the best vehicle to transfer the pollen grains.

In steps the flower.

Flowers are also used for reproduction in plants, but they have a special advantage (with a little help from their friends). The pollen grains in a flower are found on the anther. Pollen leaves the anther of one flower and needs to be transferred to the stigma of another flower. If the pollen successfully arrives and fertilizes the second flower, then seeds will eventually develop from ovules, which are in the ovary.

THE TRANSFER OF POLLEN

But how does pollen travel from flower to flower? Many of these plants have relationships with **pollinators**, meaning any animal that transfers pollen from one plant to another. Bats, hummingbirds, butterflies, moths, beetles, and bees are all pollinators. If you can get an animal to deliver pollen directly, it is much more efficient than sending pollen grains out into the wind and hoping some make it to the right place. This creates an advantage for flowering plants that work with pollinators over plants (flowering or non-flowering) that rely on wind pollination.

But why would an animal want to help a flower out by carrying pollen? Sometimes the animal is actually collecting pollen for itself, but it happens to transfer some as it makes its way around a field of flowers. Bumble bees, for example, collect pollen and bring it back to their nests to help feed their colony. They even have special structures on their back legs to collect pollen (called a pollen basket), but pollen also dusts their bodies all over as they crawl around the flower, so some will be transferred to the next flower they visit. In this way, both the bumble bee and the flower will benefit; this type of relationship, where each species benefits, is called a **mutualism**.

Another way plants encourage animals to visit their flowers is by providing **nectar**, which is a sweet, sugary substance that the animals can use as food. Bees collect nectar and turn it into honey, which we then sometimes collect from them to put on our toast. Bat pollinators are also looking for a sugary snack from flowers. The nectar encourages them to visit many flowers, and pollen that sticks to their fur gets carried from one flower to the next. Moths, ants, hummingbirds, and many other species also drink nectar from flowers.

From a plant's perspective, the partner that you interact with is important. For example, bats and birds can fly longer distances, so they may be more important for flowering plants that are spread further apart. Either way, if you're a plant that relies on animal pollination, you want to be sure some type of pollinator will visit your flowers.

What's the best color for a flower?

It depends. If your pollinators are active at night (nocturnal!), the color of the petals isn't important (many bat pollinated flowers are white). But for those animals that forage during the day, color can be an important factor in deciding where to look for nectar and pollen. Different species have different color preferences; for example, bees prefer blue and violet, while hummingbirds prefer red. To each their own! These color preferences may even help the plants, since a hummingbird that flocks to red flowers won't visit (and so won't deliver pollen to) a blue flower, which is probably the wrong species.

REWARDS AND TRICKS

Because so many flowering plants rely on pollination to reproduce, plants will compete for attention from animal partners. Offering more sugar or pollen is one way to increase visits from pollinators, and animals can actually remember the types of flowers (by sight or smell) that offer the best rewards. But plants also have other ways to compete for pollinators. One of these is to make it easy for the pollinator to find the flower, and to find the nectar once it's at the flower. This is where showy colors are important. If you have a bright color, you have a better chance to stand out in a field of flowers. Many flowers also have **nectar guides**, which are patterns that help lead the animal (like a foraging bee) to the center of the flower. We can't see some of these nectar guides, because the color (ultraviolet) is outside of our visual ability. Studies show that bees access rewards more quickly and are more likely to revisit flowers with these guides that help them find food.

Ultraviolet light (right) helps reveal the nectar guides that bees are able to see on this Mimulus flower, which we cannot see in natural light (left).

Not all flowers play fair, though. Some plants offer no nectar to pollinators, but instead trick pollinators into visiting their flowers. Some flowers resemble those that provide nectar rewards, either by sight or by smell, but don't actually provide any reward for pollinators themselves. Still others, especially in the orchid family, may trick bees by looking like a place to take shelter, or even by resembling another bee! There are likely over one thousand species of orchids that trick pollinators into visiting their flowers. The relationship between plants and pollinators is not always a mutualism; sometimes, the relationship is very one-sided!

Overall, pollination is extremely important to the plants and animals involved. This has led to many strange relationships between the partners. One of these shows the possibility for conflict between organisms, even in a mutualistic relationship, since each species (animal and plant) could increase its number of offspring (or its fitness) if it can take advantage of the other. Yucca moths lay eggs in yucca plant flowers and then pack the flower with pollen (an active transfer, not accidental like other pollinators). Then some of the developing yucca seeds are eaten by moth larvae and others develop into new yucca plants.

Yucca moth and the yucca plant (right), work together and each benefit from each other. This cooperation is referred to as mutualism.

But if the mother moth lays too many eggs and too many seeds are eaten, the yucca plant can actually drop the developing fruit, which kills the larvae and any remaining seeds. This prevents cheating by the moth and helps stabilize the mutualism.

That's it for this chapter, but remember, we are still not done with angiosperms. In our next chapter, we'll look at the incredible diversity of flowering plants.

FOUNDATIONS REVIEW

✓ The angiosperm group is a name given to flowering plants. Flowering plants represent more than 80 percent of known and described plant species. Having flowers can bring many advantages.

✓ Understanding the structure of a flower is important. Make sure to review the diagram in the chapter to remember the parts of a flower, including sepals, petals, stamen, pistil, filament, and ovary.

✓ One of the main advantages to having flowers is they help with pollination, or the reproduction of plants. Flowers can attract certain pollinators (bees, butterflies, etc.) with their bright colors and sweet nectar. The pollinator gets what it needs and wants, but in the process, pollen grains will stick to its body; when it then flies to a new flower, it brings those grains with it and pollination can occur. This is a much more efficient way to bring about pollination than depending on the wind.

St. Patrick, the Clover, and the Trinity

As we will see in the next chapter, clovers are an angiosperm, or flowering plant. That makes this space a great time to talk about St. Patrick and his use of the clover plant to explain the Holy Trinity.

St. Patrick had an adventurous life. He was born in England and lived in the fifth century. When he was a teenager, he was captured by pirates and sold as a slave in Ireland. He escaped his captors after six years, but during that time, he had grown much closer to God through prayer. He eventually returned to Ireland as a missionary after becoming a priest.

Few people in Ireland at the time were Christian, or well educated, and so it was difficult for St. Patrick to explain his faith and all the Catholic teaching he had received himself. One famous story tells how he used the clover (or shamrock) to help explain how God is one God but exists as three distinct Persons: Father, Son, and Holy Spirit. Many clover species are **trifoliate**, meaning they have leaves in sets of three, and so one plant is made up of three distinct leaves. Each leaf is fully of the plant, but each is also its own separate leaf.

Likewise, Jesus is fully God, as are the other persons in the Trinity. Three persons, one God. The use of explanations like this must have helped since many people converted to Christianity through St. Patrick's witness. This story explains why St. Patrick is often depicted with an image of a clover, or shamrock, in religious art.

St. Patrick, pray for us!

A field of wild flowers shows the amazing diversity of flowering plants.

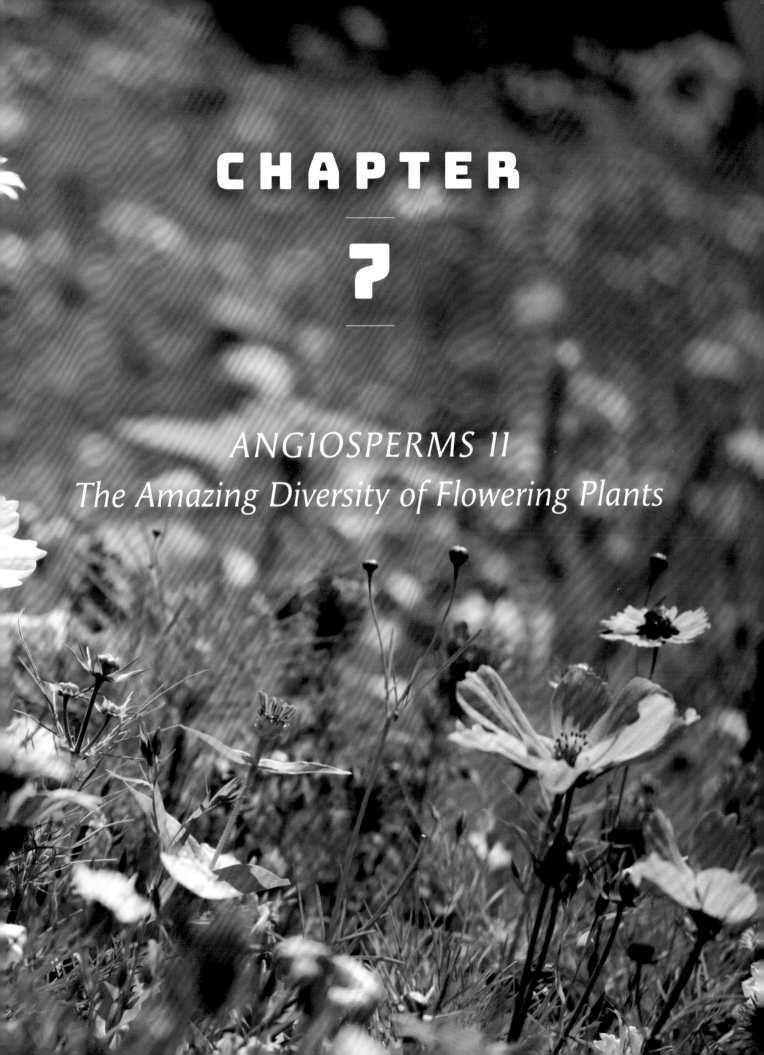

CHAPTER 7

ANGIOSPERMS II
The Amazing Diversity of Flowering Plants

Plant Fun Fact:
Magnoliids have flowers, but also have traits that match non-flowering plants. Some are used for cooking all over the world, including black pepper, cinnamon, and nutmeg. In recent years, avocados have also become very popular.

MORE ANGIOSPERMS!

It's time to continue our discussion of angiosperms. But we will shift our focus in this chapter to the wide diversity that exists within flowering plants and examine some of their features. Since there are so many families of angiosperm plants, and then species within those families, it will be hard to do this group justice in one chapter. But, to begin, we could subdivide this big group of plants into a few smaller groups.

First, we will look at flowering plants that share some traits with non-flowering plants—once thought to be an intermediate type of plant (Magnoliids). Then we'll look at the two major divisions in the angiosperms, monocots and eudicots (or dicots), and some of the traits that help us identify each.

MAGNOLIIDS

Our first group of flowering plants is the magnoliids, which includes around nine thousand species worldwide. As we've discussed, biologists like to organize and categorize things into orderly patterns. But some organisms do not quite seem to fit-in well, and scientists are always trying to learn more about them to better understand how they might fit into existing groups. The magnoliid group is one example of this. These plants do have flowers, but also possess some traits that match non-flowering plants, like how their pollen grains develop, how various structures are positioned in the flower, or how their stamens resemble gymnosperm male cones. Research continues in this area, and so maybe you can grow up to help answer these unresolved questions. In the meantime, what do we know about this group of plants?

Several traits are helpful in determining how to classify plants. This includes how many petals a flower has and the veins (or vasculature) found in the leaves. In magnoliids, the flower parts usually come in multiples of three (so three or six petals), and their leaf veins are typically branched.

Examples from the magnoliid group include plants that you may be familiar with. Many members of this group contain chemicals in their tissues that give them particular smells and tastes; thus, humans have used them for cooking all over the world, including black pepper, cinnamon, and nutmeg. Avocados are another example of the magnoliid group, as are magnolias (unsurprisingly, given their name).

MAGNOLIA FLOWER

MONOCOTS AND EUDICOTS

The next two major groups of angiosperms are the monocots and eudicots. Many species are found within these two groups, with monocots representing more than seventy thousand species and eudicots containing more than two hundred thousand species.

There are five features that help us differentiate these two big groups of plants within the flowering plants.

Cotyledons—**Cotyledon** is the name for the first tiny leaf of a baby plant. As it emerges from the seed, these plants will either have one or two leaves. This is where these plants get their name: one = mono for monocots, and two = di for eudicots.

Root systems—Monocots tend to have branching, fibrous roots, whereas eudicots have a central tap root.

Vasculature—If you cut open the stem of a flowering plant and can see how the vasculature is arranged (remember xylem and phloem?), you can tell if the plant is a eudicot or monocot. Vasculature is contained in bundles in a ring around the edge of eudicot stems. In monocots, the vascular bundles of xylem and phloem are scattered through the whole stem.

Leaf veins—Monocots have parallel veins, meaning the veins run from the stem of the leaf to the tip. Eudicot leaves have veins that branch outwards in a net-like pattern.

Flower structures—Monocot flowers have parts in multiples of three (like magnoliids), so three, six, or nine petals, for example. Eudicots have flowers with multiples of four (four, eight, etc.) or five (five, ten, etc.).

This is a great opportunity for you to get outside and identify a plant! See if you can find a flowering plant near your house or at a nearby park. Can you tell whether it is a monocot or eudicot based on its features? Even if it doesn't have flowers, chances are it is an angiosperm, so using the list of traits (other than flower structures) may still work. Next, let's look at some examples of monocot and eudicot families.

Eudicots or Dicots?

The group eudicots used to be called dicots, but the dicot group originally included many plants that we have now placed in other groups. Now the plants that are left in the group are called eu-dicot, because *eu* comes from the Greek for "good" or "true." So all the true dicots were kept in the eudicot group. Another example of science building upon itself toward better knowledge and understanding!

Above (left) is a bamboo leaf. Notice the monocot leaf veins running from the stem to the tip of the leaf. Above (right) is a mulberry leaf, an example of a eudicot leaf, with it's veins forming a net-like pattern.

EXAMPLES OF MONOCOTS AND EUDICOTS

Monocots include many grass or grass-like species, although there are others that we probably wouldn't call grass-like (pineapples!). One family of monocots is called Poaceae, but we could also call it the grass family. There are roughly twelve thousand species of grasses in the world, and they are found worldwide. Even though this family doesn't have the most species of plants, it is arguably one of the most important and most successful. More than a third of the earth's total land area is covered with grass! This plant family is vital to humans and the world food supply, because corn, wheat, rice, barley, and millet are all found in this family. Food products from these plants make up almost half of the food eaten each day.

Some of the flowers you will find in bouquets or gardens are also from the monocot group. Various species of lilies, tulips, and alstroemeria are found in several families of plants that are similar to one another. If you dissect a lily or a tulip, you'll find that they typically have six petals (a multiple of three, which fits with the monocots). Many of these species are **perennial plants**, which means the same plant will live multiple years. To do this, they have storage organs (called bulbs or corms) to store things through the winter before regrowing in the spring. The opposite of a perennial plant is an **annual plant**, which release seeds and then die, but the seeds survive the winter and then grow in the spring. These are two different strategies plants can use to make it through difficult winter conditions.

TULIP

The largest plant family is the daisy family (or Asteraceae), and it is made up of eudicot plants. There are roughly thirty thousand species in this family! Many species in this group may be familiar because they are so widespread, growing on lawns, in ditches, along roadsides, and pretty much everywhere. Dandelions, thistle, and, of course, daisies are in this family.

These species, and this family, have been so successful in growing, reproducing, and spreading to many places that some people consider these plants weeds, which is just an informal way to say a plant that you don't want around.

There are many types of green grass. Some common types are: Bermuda, Kentucky Bluegrass, Fescue, and St. Augustine. Do you know what type of grass is growing around your home?

Whether or not people want them in their yard, these flowering plants are very interesting. For example, how many flowers do you see in the picture of this sunflower? The answer may surprise you, because flowers in the Asteraceae family are unique and come in two types. Here, each petal is a "ray floret" (one type of flower; it has complete flower structure) and there are many small disc florets at the center. So each sunflower or daisy is actually made up of many, many tiny flowers!

Another large family of eudicots is the bean family, also called Fabaceae. Beans, peas, and legumes are all found in this family, and there are around nineteen thousand species total. The green beans or lima beans grown in gardens that you eat (because, of course, we all eat our vegetables) come from this family. Other species we may recognize from our neighborhoods, like clovers. But many members of this family also grow in the wild, like in tropical rainforests, for example.

One cool set of species from this group is the acacias. Some acacia plants have a mutualism with ants, but not for pollination like we read about in the last chapter. The ants protect the plant by biting animals that want to eat the

Next time you are outside exploring, look down and see if you can find this wonderful eudicot, the clover. Maybe you will find one with four leaves!

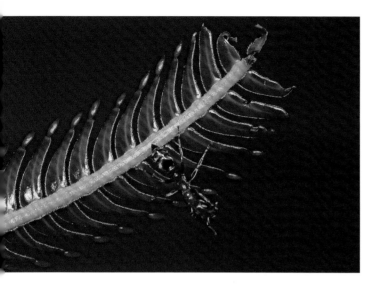

plant, and, in return, the plant provides nectar to the ants. In a weird twist, though, scientists discovered that the ants actually cannot leave the plant once they drink the nectar; they become totally reliant on that plant for food.

There are many more families and species of flowering plants, but we are out of time here, so you'll have to check out your library to learn about more these wonders of nature!

The acacias plant shares a mutualism with certain kinds of ants. The ant fends off other insects that might want to eat the plant, while the plant provides the ant with nectar.

FOUNDATIONS REVIEW

✓ Magnoliids are a common type of angiosperm, with over nine thousand species. Their flower parts usually come in multiples of three (three or six petals), and their leaf veins are typically branched. Many members of this group contain chemicals in their tissues that give them particular smells and tastes; thus, humans have used them for cooking all over the world, including black pepper, cinnamon, and nutmeg. Avocados are another example of the magnoliid group, as are magnolias (unsurprisingly, given their name).

✓ The next two major groups of angiosperms are the monocots and eudicots. Many species are found within these two groups, with monocots representing more than seventy thousand species and eudicots containing more than two hundred thousand species. There are five features that help us differentiate these two big groups of plants within the flowering plants: cotyledons, vasculature, root systems, leaf veins, and flower structures. Monocots include many grass or grass-like species, as well as some of the flowers you will find in bouquets or gardens, while eudicots include those plants found in the bean family.

✓ Many of these species are perennial plants, which means the same plant will live multiple years. To do this, they have storage organs (called bulbs or corms) to store things through the winter before regrowing in the spring. The opposite of a perennial plant is an annual plant, which release seeds and then die, but the seeds survive the winter and then grow in the spring. These are two different strategies plants can use to make it through difficult winter conditions.

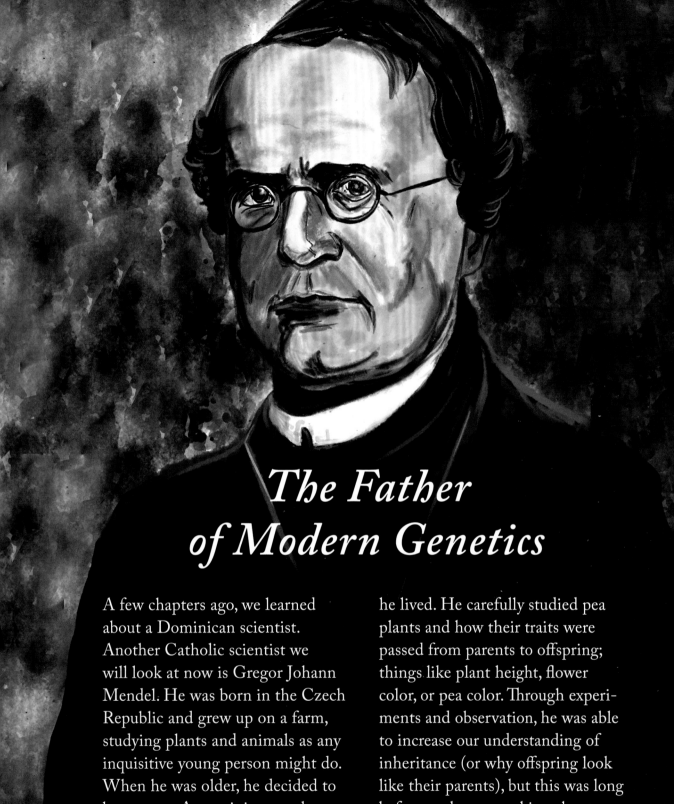

The Father of Modern Genetics

A few chapters ago, we learned about a Dominican scientist. Another Catholic scientist we will look at now is Gregor Johann Mendel. He was born in the Czech Republic and grew up on a farm, studying plants and animals as any inquisitive young person might do. When he was older, he decided to become an Augustinian monk.

As a part of his life as a monk, he got the opportunity to receive an education and study many topics. His most famous work was conducted in the garden of the monastery where he lived. He carefully studied pea plants and how their traits were passed from parents to offspring; things like plant height, flower color, or pea color. Through experiments and observation, he was able to increase our understanding of inheritance (or why offspring look like their parents), but this was long before we knew anything about genes or DNA or could look inside cells to see how it actually worked.

For his discoveries, Gregor Mendel is known as the "father of modern genetics."

The lion's mane mushroom is certainly an example of an odd-looking plant! Or is it a plant at all? Decidedly not! Although mushrooms are plantlike, they are actually a type of fungi.

CHAPTER 8

NON-PLANTS AND STRANGE PLANTS

PLANT ODDITIES

Up to this point, we have looked at several major groups of plants, characterized by some of the major features and key traits that help us categorize these organisms. But we have focused specifically on land plants and those that fit in well with other plants.

In this chapter, we will instead focus on some oddities in the plant world, like plants that eat meat, or ones that don't use photosynthesis. We'll also take a closer look at which organisms are plants and which are not. This will include fungi and some seaweeds, not because they are plants—they aren't—but because they are often mistakenly described as being plants or related to plants. So let's take a look at a few of these odd balls!

ALGAE, SEAWEED, FUNGI, AND MORE!

You may have noticed that our examples from past chapters were plants that live on land. Most plants are contained in a group called Embryophytes, which contains all land plants, but doesn't include some other organisms that are also defined as plants (like algae). In this textbook, and in the companion *Animals* book, we often discuss various groups of similar organisms. These are divided into layered categories, with each subcategory being more specific and containing fewer, more closely related species. The groups, from largest to smallest, are:

Domain > Kingdom > Phylum > Class > Order > Family > Genus > Species

The exact names of the groups aren't important here, but notice that Kingdom is the second largest grouping and so contains many diverse organisms. For example, all animals, from lions to ladybugs, are in the animal kingdom. The land plants we have discussed are in the **kingdom Plantae** (or Plant kingdom), but so are some organisms we would not immediately recognize as plants.

BLACK FUNGUS

Plant-like organisms like algae and seaweed that live in water have been a puzzle for scientists. When you find green things growing in a lake or at the ocean, it may take a plant biologist to say whether it is in the plant kingdom or not. For example, algae are a complicated group of aquatic, photosynthetic organisms. Some are quite small, having only a single cell. But other species, like giant kelp, can grow more than one hundred feet upward from the ocean floor! Because they can conduct photosynthesis, and share some other similarities to the land plants we have discussed, all algae were once considered to be plants.

But with advances in technology, like better microscopes to look very closely at each cell, scientists realized that algae were quite a diverse group. It turns out that some algae more closely resemble plants than others. Now we have several algae groups (brown, green, and red), including some that are typically still kept in the plant kingdom (the green algae). But other algae species, including the giant kelp (a brown algae), are no longer considered plants because they lack important plant features.

Plant Fun Fact:
Giant kelp can grow more than one hundred feet upward from the ocean floor!

Another group of organisms once considered to be plants is fungi. Upon closer inspection, though, fungi are even less plant-like than any algae or seaweed. In fact, fungi are more closely related to animals than they are to plants. How can this be true? Well, there are several key features of plants that fungi simply do not share. Fungi, like mushrooms, are not photosynthetic. They get their food mostly by breaking down and absorbing nutrients from dead and decaying things. Fungi also use something called **chitin** (which we discussed in chapter 1) for structural support, and chitin is found in the animal kingdom (in the exoskeleton of insects, or some fish scales). Plants use a completely different molecule for support—**cellulose** (also discussed in chapter 1)—which is not found in animals.

What are some examples of fungi? Mold, mushrooms, and yeasts are all fungi. Some are important to humans (like baker's yeast for bread-making) while others are harmful, causing disease or infecting our skin. But overall, both plants and fungi are critical to the ecosystem—plants for photosynthesis and fungi for recycling nutrients for organisms to use again.

With some of these strange plants (and non-plants) behind us, let's take a look at plants that eat animals and steal their food from other plants!

Puffball fungus got its name because clouds of brown dust-like spores are emitted when the mature fruitbody bursts or is impacted.

CARNIVOROUS PLANTS

As we know by now, plants need nutrients (like nitrogen and phosphorous) to grow, but sometimes the soil they grow in does not have much of these things. So how can a plant get enough nutrients to grow and thrive, since they can't move around like animals to go find some?

Well, speaking of animals, it turns out they are a good source of nitrogen and phosphorous. Some plants seem to have developed devious ways to capture animals to help provide them with these critical nutrients. Because they eat

Mushrooms Larger than a Blue Whale?

The largest organism on earth is a fungus. A honey mushroom, like other mushroom species, mostly grows underground—the mushrooms we see are sent to the surface because that's how the mushroom reproduces. But much of the organism remains hidden underground. One particular honey fungus in Oregon is estimated to be more than 2 miles (yes miles!) across, and cover more than 3.5 square miles total. This makes it much larger than blue whales, dinosaurs, or any other giant organism!

animal tissue, this makes them **carnivorous plants**. Despite this odd lifestyle, they are still plants because they have roots and leaves and flowers and other plant characteristics. Many of them also still conduct photosynthesis, but the carnivory gives them an added boost of nutrition.

Sundew plants have sticky nectar that attracts insects, and the leaf coils up and digests the insects that get stuck there. Pitcher plants, meanwhile, have

Drosera capensis, commonly known as the Cape sundew, captures a fly in its sticky nectar.

pools of liquid in specialized leaves (like a pitcher of water). The sides of the pitcher are usually slippery, so insects that crawl on the plant may slide into the bottom. Once it has fallen in and drowned, the plant digests the insect. Other pitcher plants have turned to another source of nitrogen and phosphorous: animal dung! A particular pitcher plant species, *Nepenthes hemsleyana,* has a mutualism with bats. The plant has specialized leaf structures to help echolocating bats find the plants. In turn, the bats use these pitchers as a daytime roost. While they hang upside down in the plant to sleep, they also excrete feces, which the plant digests as a source of nutrients. Several other pitcher plant species offer sugary nectar to small mammals who visit the plant to lick up the nectar. The interesting thing is that the nectar is positioned so that the animal is sitting right over the top of the pitcher while it eats, so that it's like a tiny toilet for shrews (a mole-like animal) that offers them a snack!

PITCHER PLANT

NON-PHOTOSYNTHETIC PLANTS

Carnivorous plants still conduct photosynthesis, but there are some plants that do not. Instead of photosynthesis, they break down decaying material, like fungi, or even steal food from other organisms. Many of these **non-photosynthetic plants** are **parasites**; the term *parasite* means any organism that feeds off of another organism. Fungi and plants often interact with one another, trading

Monotropa uniflora (shown above) is also know as the Ghost plant and grows in regions of Asia, North America, and northern South America.

nutrients with each other to the benefit of both organisms (in fact, without fungi, many crop plants we rely on would not produce nearly as much food). But some parasitic plants take advantage of this trading system, taking nutrients from fungi but not offering anything in return (like the beautiful pink plant *Monotropa uniflora*). Other parasites get water and nutrients by directly absorbing it from a host plant. They do this by growing root-like haustoria into the xylem and phloem of another plant. Then, as the plant uses its vascular transport system, some of the nutrients are stolen by the parasite. While some parasitic plants get all of their sugars from other plants, others still conduct some photosynthesis (like mistletoe).

LOOKING AT THE WHOLE PICTURE

This chapter ended up as a hodge-podge of odd plants, plant relatives, and non-plants. But, in addition to just being fascinating, these examples can help us learn more about how scientists group organisms together. We cannot just look at one trait, or even a few, since some plants don't conduct photosynthesis, and some non-plants may look like plants on the outside. We need to look at the whole picture to help us understand and categorize living things.

While we spent a lot of time in this chapter discussing what plants eat, let's change focus in the next chapter and discuss what plants we eat.

FOUNDATIONS REVIEW

- ✓ Often there are organisms that look likes plants and get called plants, but actually aren't. These include certain kinds of algae, fungi, and seaweed. There are also other oddities in the plant word, like carnivorous plants and non-photosynthetic plants.

- ✓ For example, fungi are actually more closely related to animals than plants. They get their food mostly by breaking down and absorbing nutrients from dead and decaying things. Fungi also use something called chitin for structural support, and chitin is found in the animal kingdom (in the exoskeleton of insects, or some fish scales). Plants use a completely different molecule for support—cellulose—which is not found in animals.

- ✓ Some of the most interesting plants in the world are carnivorous plants, which means they feed on animal tissue. Sundew and pitcher plants have sticky nectar or pools of liquid that attract insects which they then capture and eat, or some cases these plants engage in a mutualism with animals, giving up their nectar in exchange for eating the animal's dung.

St. Francis Xavier

One saint who may have had a chance to see many strange and diverse plants and animals is St. Francis Xavier. He was born in what is now modern-day Spain, but he spent his life traveling the world to preach the Gospel. He was one of the first Christian missionaries to visit many places in Asia, spending time in India and on islands in Southeast Asia and off the shore of China, and also in Japan. So perhaps he saw the extremely unique looking black bat plant in China, or the Japanese Snake Gourd that climbs and tangles on other plants and only blooms at night, or maybe the "corpse flower" of Southeast Asia that smells like decaying meat to attract flies for pollination.

St. Francis Xavier traveled thousands of miles in his lifetime, by ship and on foot, seeking out those who had not heard about Christ and his love. There must have been many challenges that he was faced with during his journeys: from storms and long days of travel to learning new languages and customs. But through his evangelization, Christianity continued to spread wherever he visited.

St. Francis Xavier's travels were focused on his mission of teaching people about Jesus, but I do hope that he also got a chance to see some of the wonders of creation while visiting these many places. I don't know if he ever ventured into the forests of Borneo when he visited, but maybe St. Francis Xavier even got to see the pitcher plant that doubles as a shrew toilet with a feeding station.

Who knew plants could be so delicious? Fruits and vegetables make up a large portion of our diet, making farming and agriculture one of the biggest businesses in the world.

CHAPTER 9

PLANTS AS FOOD

Plant Fun Fact: Perhaps surprisingly, pumpkins and avocados are considered fruits.

PLANTS SUSTAIN ALL LIFE (IN ONE WAY OR ANOTHER)

Thus far in our book, we have focused on different types of plants and their features. Most of the rest of this textbook will look at different ways that plants are important to ecosystems, including how much we depend on them. Specifically in this chapter, we will explore how we, and other animals, use plants for food.

Many animals eat plants. As we spoke about earlier, plants form the base of the food chain because they can convert sunlight into food energy (like sugars), which they store in their tissues to use later on for growth, maintenance, or reproduction. All that energy provides good nutrition and food energy for animals.

If you read the *Animals* textbook in our series, you might remember that an animal that eats plants is an herbivore. If it eats meat and plants (like we do), then it is an omnivore. But even animals that don't eat plants and only eat meat (carnivores) still rely on plants, because the animals they eat can only grow and reproduce by eating plants. So in some form or another, plants help sustain all life on earth.

But when an animal eats a plant, what part does it eat? Well, it depends. Let's take a closer look at the interactions between animals and plants by looking at what animals eat.

LEAVES, ROOTS, RHIZOMES, AND TUBERS

One part of the plant that some animals eat are the leaves. This presents a problem for plants, because these are the factories that produce sugar from light energy through photosynthesis. Perhaps for this reason, many leaves (and other above ground plant tissues, like stems) are not very nutritious—as soon as sugars are produced, the plant can send this food to be stored elsewhere.

The nutrition in the plant parts has effects on herbivores. For example, scientists have shown that monkeys that eat more leaves in their diet (instead of fruits, insects, or other things) need to spend more time eating and digesting,

and so have less time for other activities. The same is true for animals that eat grass (which is mostly leaf and stem); large herbivores like horses may graze in a pasture for as many as twelve hours per day! But some leaves are more nutritious, and those tend to be the ones we humans grow and eat. Things like spinach, kale, and cabbage are all leaves that we eat.

Some foods we eat come from plant tissue underground, like roots, rhizomes, and tubers. We know about roots, but what is a tuber or a rhizome? **Rhizomes** are stems that grow outward away from the plant underground. They are structured like stems, and branch off the main stem, so are not roots, even though they are underground. Some rhizomes we use in cooking, include ginger, turmeric, and arrowroot. A **tuber**, meanwhile, is a swelling of plant tissue that is used to store water, starches, sugars, or other things the plant needs. This could either be a tuberous (swollen) underground stem, like potatoes and yams, or a tuberous root, like sweet potatoes.

Last of all, some roots themselves are food for people and animals. Beets, parsnips, radishes, and carrots are all taproots. All of these different examples come from different plant tissue, but each stores food for the plant. Why do you think these are found below the ground? One answer might be that it makes it more difficult for animals to eat the food the plant has stored up for itself. Unfortunately for the plant, we humans have shovels and farm equipment, so we can get the food stored below ground. Animals that eat these foods often have their own digging equipment as well, like strong digging claws. An easier to access source of food, found above ground, are the fruits of the plant.

RED BEET

FRUITS

What is a fruit, anyway?

Well, there is some confusion over this term, because we often talk about fruits versus vegetables when discussing nutrition and eating a balanced diet. In nutritional or dietary terms, fruits are usually the sweeter food items that grow on plants, like apples, oranges, peaches, or pears. But in **botany**, or the study of plants, a **fruit** has a specific definition: anything that contains seeds and develops from the ovary of the flower. This means that many things we call vegetables are actually fruits—cucumbers, tomatoes, and green peppers all develop from flowers and have seeds. Perhaps more surprisingly, eggplant, olives, pumpkins, and even wheat (and other grains) are all fruits. Or at least this is true from the perspective of a plant scientist.

Interestingly, there are other terms that we commonly use to describe foods that come from plants, but that are not used in their scientific, botanical meaning. Berries are a good example of this. If you asked a friend or your parents to name a berry, chances are they might list things like strawberries, blueberries, or raspberries. The problem with that is only one of those (blueberry) fits the definition of a berry. A **berry**, in terms of plant science, is a simple fruit that develops from a single flower with a single ovary, and that has multiple seeds (not a single pit or stone like cherries). Fruits that fit this definition, based on how they develop, include grapes, tomatoes, cucumbers, bananas, and watermelon. Yes, watermelon!

Most people would see an assortment of berries here, but not all of them are actually berries. Can you name these and determine which are truly berries and which are not?

The confusion between common language and botanical language has even more odd examples. One of these is that not only is a strawberry not a berry, it isn't even a fruit! Well, sometimes it is called an "accessory fruit," but the strawberry itself does not develop from a flower; instead, it develops from the **receptacle** (this is the part where sepals and petals attach). The seeds on the outside can be considered fruits, though. Fruits develop from the flower ovary, and each seed on the outside does this.

Each of these berries, like many other fruits, has three layers that you could try to identify yourself. These layers are the **exocarp**, **mesocarp**, and **endocarp**. These are scientific words to say the outside (exo=outside), middle (meso), and inside (endo) body of the fruit. You can dissect a fruit like an apple to view these layers. The skin of the apple is the exocarp. On some fruits, this is thicker and so we don't eat it (like banana peels). The exocarp can help protect the fruit and seeds. The mesocarp is the fleshy middle of the fruit, between the skin and core of the apple. This part usually offers the most nutrition to humans or animals that might want to eat the fruit. Then finally, the very inside, like the core of the apple, is the endocarp. The endocarp can provide more protection for the seeds, but also serves to help the parent plant tissue (which the fruit develops from) and the offspring (the seeds). Just like a mother helps provide nutrition to a baby in her womb through an umbilical cord, the endocarp does this with the developing seeds through a structure called the **funiculus**. Overall, a major biological purpose of a fruit is to help seeds get a good start in life.

One of the ways fruits help parent plants and seeds is to work with animals. Some plants try to avoid having their fruits and seeds eaten (we will talk more about that in a chapter about plant defenses). But others really don't mind if animals eat their fruits.

This may be part of why many fruits have bright colors—to actually attract herbivores to eat them. Another important job for fruits is to help young plants move farther away from the parent plant. When we talked about plants with seeds (in the gymnosperm chapter), we mentioned that animals can help plants disperse their seeds—they eat the fruit at one place, digest it, and then excrete the seeds out in another place, sometimes miles away. One fascinating example of this is tapirs in the Amazon rainforest. More than twenty species of plant seeds have been recovered in tapir dung. Furthermore, because of

Animals like this South American tapir help plants disperse their seeds in areas that are miles apart.

Plant Fun Fact:
Each pineapple plant only produces one pineapple!

where they spend their time, tapirs were found to deposit more seeds in poor forest habitats, like places where a fire has come through. In this way, they can actually help rainforests recover from a fire or some other situation where it was destroyed. Any help regenerating rainforests is good thing, because they are home to so many diverse plants and animals, and also help provide us with the oxygen we need to breathe.

After seeing how plants are gathered and consumed as food here, we will shift our focus in the next chapter to discovering how plants can actually defend themselves from being eaten.

Anthropomorphism

Arguably, plants don't "mind" anything, since they aren't thinking or making decisions like we do. Using language like this where we talk about animals or plants almost with human like qualities is called **anthropomorphism**. Usually this isn't too much of a problem, as long as we are keeping in mind that we are distinctly different from nature—primarily because of our immortal soul!

FOUNDATIONS REVIEW

- ✓ In some way or another, plants sustain all life on earth. Many animals eat them, including us, and even animals who only eat meat depend on plants because the animals they eat survive on a diet of plants. Furthermore, plants are how energy enters the food chain, since they can turn sunlight into nutritious sugars.

- ✓ Leaves are one of the most common parts of a plant eaten by animals, but certain roots are also edible, like carrots. Rhizomes (stems that grow underground) and tubers (swollen plant tissue) are also sources of food. These include tuberous (swollen) underground stems, like potatoes and yams, or a tuberous root, like sweet potatoes.

- ✓ Fruits are a confusing kind of plant because they are often mislabeled in common language. Technically speaking, a fruit is anything that contains seeds and develops from the ovary of the flower. This means that many things we call vegetables are actually fruits—cucumbers, tomatoes, and green peppers all develop from flowers and have seeds. Berries are another example of mislabeled plants; for example, a strawberry is not really a berry, and in fact is not even a fruit! This confusion stems from differences arising between common speech and technical scientific terminology.

Fruits of the Holy Spirit

Plants use fruits to help their offspring—the seeds—ultimately be more successful in getting their start. A plant that bears more fruit is more successful. This is true from the plant's perspective as well as from the gardener's perspective. Perhaps this is why fruits are often used to describe something that is the successful result of something else, such as "the fruits of our labor" being the results of our good work.

One placc this is used in the Bible is when Paul talks about the fruits of the Spirit. These are the results of when the Holy Spirit is working within us.

The Fruits of the Holy Spirit are:

Love
Joy
Peace
Patience
Kindness
Goodness
Faithfulness
Gentleness
Self-control

Can you define what each of these are and give examples of them in your daily life?

Prickly pear cactus blooming, with an impressive natural defense against enemies.

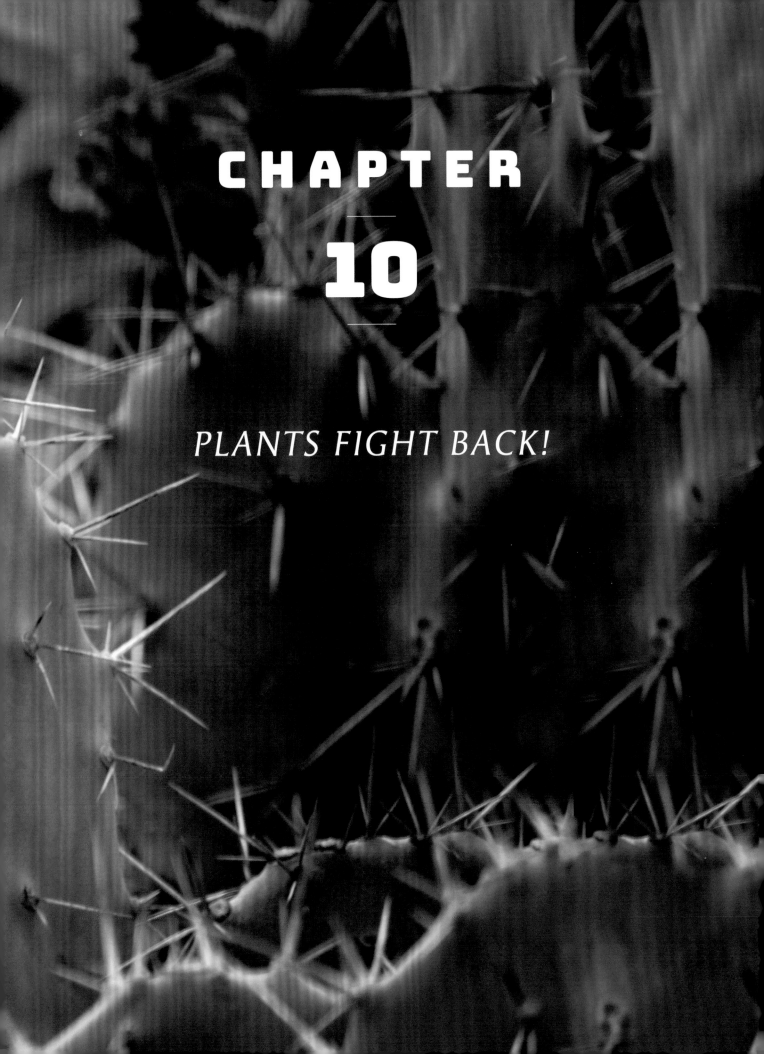

CHAPTER 10

PLANTS FIGHT BACK!

Plant Fun Fact:
Some cactus flowers bloom for only a day, while others last for weeks. Some only bloom at night!

CHALLENGES TO A STATIONARY LIFE

In the last chapter, we looked at how humans and animals eat plants, or at least certain parts of plants. Sometimes this happens within a partnership, like when animals eat fruit to help disperse seeds. But usually, as you can imagine, being eaten is a bad thing if you are a plant.

Of course, being eaten is an ordinary thing in the natural world. Animals are eaten by other animals all the time, but they at least have the advantage of being able to run away from their predators, unlike plants, which are obviously stationary. There are also other challenges with being stuck in one place. For example, seeds grow where they germinate, but they cannot leave if there is a lot of competition from other plants in that place. By "competition" we mean perhaps that there is not enough room for root systems to grow and become strong, because they will run up against other plants' root systems, or that there is not enough water to be soaked up, among other competitive issues.

In this chapter, we'll look at how plants avoid being eaten, as well as some of the ways they interact with each other and sometimes even fight for resources.

CHEMICAL DEFENSES

One way that plants defend themselves is with chemicals. If part of the plant, like a leaf, is nutritious and delicious to animals that want to eat it, the plant could add something to make it taste bad or even make it poisonous. Because they don't have legs to run away from predators, members of the plant kingdom have traits that give them quite a mastery over chemistry. You can experience this yourself with some plants that we eat. Horseradish plants, for example, have two chemicals that, if kept separate, don't really pose any danger to organisms. But when an herbivore (including a human) chews up the plant, it damages the cells. Little containers that hold the chemicals burst, mixing the

Horseradish roots (right) can be made into a spicy puree that has a kick (below).

chemicals together. A chemical reaction then follows, and a new chemical (allyl isothiocyanate) is formed; this irritates eyes and lungs and tastes terrible (a *small amount* of this chemical gives horseradish sauce its distinct kick).

Another example of this comes from spicy peppers, like jalapeños. Peppers have different amounts of a chemical called **capsaicin** that causes a sensation of pain when it binds to some of our taste receptors (jalapeños have some, ghost peppers have a lot). It turns out, though, that only mammals experience pain from capsaicin, because we have the right receptors. Birds, on the other hand, do not have receptors for the capsaicin to bind to, and so they can eat spicy peppers with no issues.

Some scientists have hypothesized that having capsaicin, rather than another chemical that might affect all animals, may provide advantages to the plant. Why would it be beneficial for birds to eat the peppers rather than a mammal? Well, there may be a few reasons. First, more mammals may be seed predators (meaning they eat the seed, like mice), and plants don't want to lose their seeds. Also, birds could eat the seed and disperse it quite far from the parent plant, since, after all, they can fly. Having their seeds travel farther means more room for growth.

JALAPEÑOS

Not all chemicals are used to deter herbivores, though. For example, some plants release chemicals into the air when they are being eaten by caterpillars. Why would a plant do this? It may function as an alarm or a call for help. There are animals that like to eat caterpillars, so if a plant can attract those animals, it may have a way to get rid of the little chewing machines. Studies have shown that parasitoid wasps can detect and follow the chemical signal (which is basically like a smell). These wasps then capture caterpillars and fly them away to bury them underground. But they bury them only after laying eggs onto them; their young, when they hatch, will then eat the caterpillar. Quite a gruesome end for the caterpillar, but a happier ending for the plant! This type of defense is called an indirect defense (or biotic defense) because it relies on the help of another organism.

STRUCTURAL DEFENSES

In addition to chemical defenses, plants also have structural defenses. These are defenses built into the plant itself. To see one example, we would need to zoom in to look at individual cells in the plant. Each plant cell has a **cell wall**, made up of cellulose and other structural components. This structure helps protect against plant diseases—yes, plants can get sick too! The cellulose is also difficult to digest, which makes it double as a protection against herbivores. Most animals cannot eat wood because of its high content of indigestible parts. Like many of these defenses, making plant tissues difficult to digest protects against many generalist herbivores. But there are specialists that have ways to get around these defenses. Termites have symbiotic bacteria that live in their guts and help them break down and successfully digest wood.

We do not need a microscope to see other structural defenses. Thorns, spines, and prickles can be an effective way for a plant to prevent herbivores from eating their valuable tissues. All of these structures are sharp, poky ways to keep herbivores away—and it would certainly be unpleasant to try to eat a cactus. Each type of spike is defined by what it is made up of. **Thorns** are parts of the stem or branches that are modified into spikes. **Spines**, like on a cactus, are modified leaves. **Prickles** are formed from the outer surface of the plant tissue (the epidermis). You may have thought that roses have thorns, but, technically, they have prickles.

From left to right; large honey locust thorns, green cactus spines, and the rose stem prickles.

PLANT COMPETITION

So far, we have looked at some of the ways plants protect themselves from animals, including asking for help from the enemies of their enemies. But plants also interact with each other, including competing for limited resources (water, nutrients, sunlight, etc.).

Don't Mess with the Cactus!

Since the spines on a cactus are technically modified leaves, this means cactuses don't have a lot of photosynthesis going on, since their leaves are used for protection instead. But what color is the rest of the cactus? Green! This means cactuses have chloroplasts stored in other places to help with photosynthesis. Small, pointy leaves also helps a cactus grow in dry deserts, because this modified leaf shape does not lose as much water.

Sunlight is a critical resource for plants since they build their food using light energy. Plants have the ability to sense light levels, and so often grow towards the light. It turns out that plants can also determine how much competition is around and respond accordingly. Scientists have shown that when plants are in full sunlight, they grow more branches and leaves outward. But if it grows in a shady area, chemical signals in the plant's body tell it to grow longer and taller to try to reach more sunlight. Sometimes plants also give up; if they are stuck in the shade, they may shift to producing seeds instead of growing. Hopefully then some of their offspring can end up in a sunnier spot.

There are other ways plants compete with neighbors for sunlight and other resources. Some actually release chemicals that prevent other plants from growing nearby. If no competition can grow, all water and nutrients in the area are available. Black walnut trees are a good example of this form of chemical warfare. Parts of the black walnut tree release chemicals, which prevents plants from growing under and around the tree. One of these chemicals, called juglone, is especially present in the buds, nut hulls, and roots. The chemical may both prevent animals from eating plant parts (it can be toxic to animals) and also prevent plant growth once it is in the soil. My gardening tip for all of you is to avoid planting a garden under a black walnut tree!

One last fascinating, though extreme, example of plant competition is to burn up the competition and provide a healthy start for young seedlings. We heard

This black walnut in Oregon has been successfully "defending its territory," as we can see that no other plants (besides grass) or smaller trees have grown up around it.

Plant Fun Fact:
Even though fires can devastate plants as the flames spread across a given terrain, some species can benefit, as they grow back quicker and benefit from the lack of other plants competing for resources in the wake of the scorched earth.

about some pine tree seeds that germinate and grow after wildfires; this means that they start growing when there is less competition for resources. Some of these plants have combustible chemicals which would help spread the fire further. There are even stories of rockrose plants that spontaneously burst into flame on extremely hot days! I don't think this has been confirmed by scientists, but we do know that seeds from these plants are fire resistant and will germinate after fires.

Overall, these are only a few examples of how plants interact with each other and with animals—both helpful and harmful animals. It is clear that even though they cannot move around, plants play an active role in their community and can fight for survival in all sorts of creative ways. We'll continue this discussion in our next chapter in regards to how plants benefit communities and ecosystems.

FOUNDATIONS REVIEW

✓ One of the primary challenges to being a plant is defending itself against animals and the elements since they are stationary. But plants have come up with several different ways to survive the various threats they face.

✓ Two specific methods of defense are through chemicals and structure. Some plants will release a chemical that makes it taste bad or can even be poisonous, while others develop thorns, spikes, and prickles as structural defenses. (No one wants to eat a thorn, not even a hungry animal!)

✓ Plants also compete with each other when it comes to getting the resources they need. Some have learned to grow a certain way based on where they are planted so that they receive more sunlight than their neighboring plants, and the black walnut tree will actually release harmful chemicals that prevent other plants from growing up and stealing its sunlight and water.

The Providential Care of God

We see that plants are not defenseless against their enemies. In nature, God established a system that allows plants (and animals) to confront the dangers that they face.

Much more so, though, God actively loves and cares for us. After all, in the Gospel of Luke, we read: "Are not five sparrows sold for two pennies? And not one of them is forgotten before God. Why, even the hairs of your head are all numbered. Fear not; you are of more value than many sparrows" (Lk 12:6–7).

Plants, nor sparrows, have escaped the notice of God. But each of us are more valuable than many sparrows, as the verse says, as we are made in the image and likeness of God. Thus, we can be confident that he watches over us—all the hairs on our heads have been counted! In good times or times of trouble, we can turn to God in prayer and know that he will protect and care for us.

A lush rainforest teems with plant life amidst a morning fog.

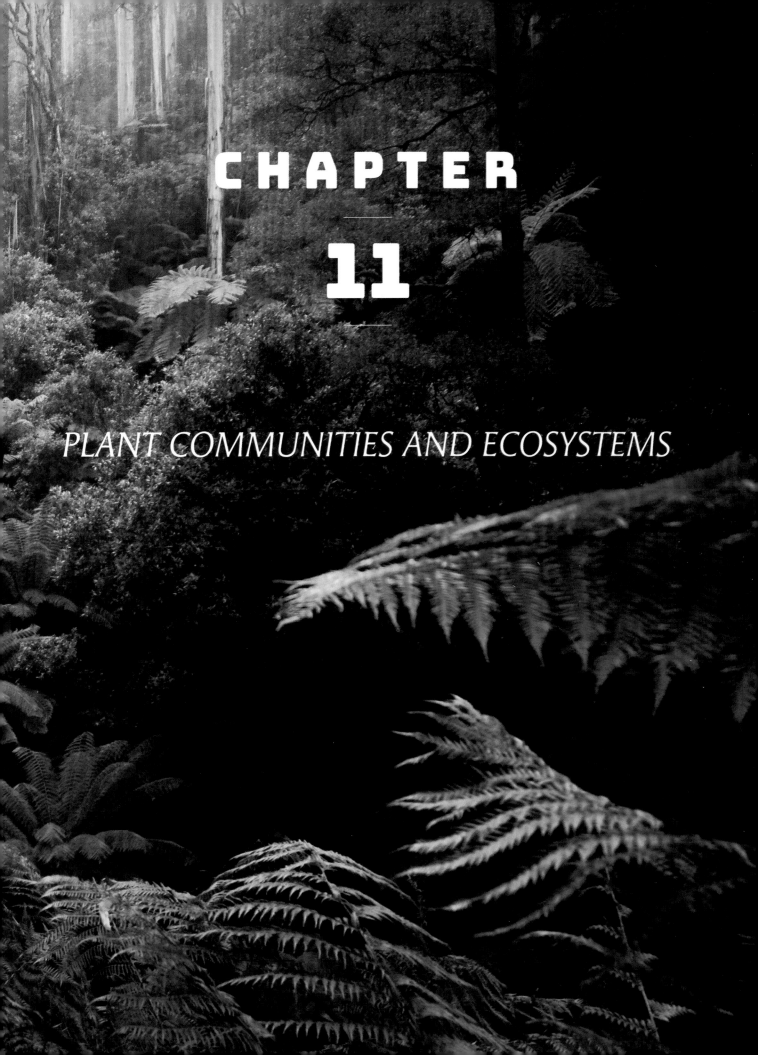

CHAPTER 11

PLANT COMMUNITIES AND ECOSYSTEMS

HOW PLANTS AFFECT BIOLOGICAL COMMUNITIES

In the last chapter, we looked at some of the ways plants interact with other animals and plants around them. All of these interactions add up over time, sort of like how in cartoons a snowball gets bigger and bigger as it rolls down a hill, and plants end up having a big effect on the communities around them. The term **community**, when used in biology, simply means all the organisms in an area that interact together, sort of like how your church is a community of humans that interacts together every Sunday in a particular location. A biological community includes plants, animals, fungi, and even bacteria—all of these living things have effects on each other. A textbook focused on ecology would spend a few chapters talking about communities and species interactions, but here we will just take a quick look at how plants, as the stars of our book, affect biological communities.

THE SUCCESSION OF PLANT LIFE

I started out by saying that the interactions of plants with other species will add up over time, and so the first thing we'll look at is a long-term process called

succession. This refers to the change in a community over time, but in ways that we can predict based on the features of different plants and their interactions with each other. Succession, in the ordinary use of the word, means when one thing follows another. A good visual of this might be a string of dominos that are stacked up beside each other; tip over one, and the next tips, and the next, and so on.

To see what this means in the world of plants, imagine a patch of bare, rocky ground—no plants or other organisms in sight. We would expect that small plants that grow quickly, and can disperse their seeds very far from parent plants, would start growing there within a year or maybe two (think small weeds, grasses, or moss). But it turns out that some of these first species to show up are not as good at competing with other plants. They don't grow tall, and so might be shaded out once taller plants arrive. Other medium-sized plants might arrive next, and eventually tree species might take over. But it would be decades or centuries before this patch becomes a mature forest.

There are two main types of succession: primary and secondary. **Primary succession** is starting from nothing—no plants or organisms, but also limited or no soil to grow in. A good example would be any new island to emerge from the ocean. As an underwater volcano spits out lava, it hardens to rock and can form an island (this would be the first domino falling in the succession). The lava rock is not good for growing things, but mosses and other small plants may be able to eke out a living there. Eventually, enough mosses and other pioneers in this new place will begin to grow, then die and turn into soil. (More dominos are falling!) Once some soil is formed, other, taller plants could colonize the area. Eventually, over *thousands* of years, an island could turn into a tropical paradise like Hawaii.

The formation of an island through underwater volcanic eruptions is an example of primary succession.

Secondary succession takes less time, since it doesn't start from zero. **Secondary succession** is the process of plant life growing back after a disturbance, like a wildfire. Another example is when agricultural fields, where farmers may have cultivated crops, are abandoned and left to turn back into wild areas.

Secondary succession occurs at a much faster rate than primary succession.

Amazing Lichen

Another important player in the process of primary succession is lichen. **Lichen** is a crusty or sometimes leaf-like, flat growing organism that you can find on bare rocks and trees. Lichen is amazing, not only because it can grow practically anywhere, but because it is a partnership between fungus and algae. The fungus is the part you see, but nestled inside is a photosynthetic algal partner. The fungus provides a safe place to live, and the algae provides nutrition to both partners from photosynthesis.

SURVIVING IN DIFFERENT HABITATS

From the example of primary succession, it should be clear that some plants can grow in conditions that others cannot. A giant sequoia tree could never grow without soil on lava rock, but moss could. The presence or absence of good soil is a pretty extreme example, but there are many other factors that plants need to deal with which determine whether they can grow in an area. The amount of nitrogen or other nutrients, the amount of rainfall every year, and the average temperature in summer or winter are all factors that could determine the success of plants in a particular area.

As another extreme example, cactuses can grow deserts with high heat and low amounts of rainfall per year. To do this, the cactus has special adaptations, like how we previously mentioned its spiny leaves that help reduce water loss. Other desert plants have thick waxy layers to prevent water loss, or may have extremely long root systems to help find water. (Mesquite trees can grow root systems over eighty feet long!)

None of these adaptations, though, would be quite right for success in places with lots of rain, like the rainforest. Rainforests, aptly named for all the rain they receive, can get more than fifty times more rainfall than a desert. This leads to a lot of plants growing, which means a lot of competition for light and resources. So competitive traits might be more important in this habitat. For example, plants often grow quickly, and trees grow straight upwards to reach the light at the top of the forest.

Because the adaptations of a plant are features that are directly suited for the environment in which they live, we find distinct sets of plants in different areas on earth. For example, the features found in desert plants allow them to live in dry areas, so the same types of plants are found in desert-like regions across the southwest United States. Similar looking plants, with similar traits, are found in other desert areas around the globe. The same is true for animals, in part because they depend on plants for food (or for plants to be the food of their food!). These distinct sets of types of plants and animals that are found in similar habitat types are called **biomes**. A desert is one biome type. Others include tropical rainforests, savannas, deciduous forests, and tundras.

Biomes look different from one another because there are differences in weather and climate. Some areas have cold winters and hot summers (temperate deciduous forests), but others are hot most of the time and have rainy and dry seasons (savanna). The plants in each biome are uniquely fitted to the conditions in that area. So climate affects plants, and then plants present in an area end up determining a lot about the habitats and food available to animals. Overall, we end up with unique sets of organisms making up the communities that we see in different parts of the world.

Landscapes vary greatly depending on their habitats. The tropical rainforest, desert, forest landscapes, and savanna all create unique plant life suited for each environment.

PHENOLOGY

Importantly, this also means that as the climate changes, we may see shifts in the plants and animals present in an area. If animals and plants respond differently to changing climates, this could also end up disrupting communities or changing biomes. As an example, animals, or at least large ones, may be able to move to better climates more easily. This means animals and plants that are typically found together may not be in the same places any more. Or plants and animals may start to change when they do things during the year because of warmer spring or fall seasons.

For example, a plant may start flowering earlier with warmer climates, but bees may not be active at those times. The seasonal timing of events like flowering, when fruits set, and other natural events happening is called **phenology**. As these changes continue to happen over time, scientists will continue to study them because of the important links between climate, plants, and animals.

BIOMASS AND PRODUCTIVITY

Plants are important to their communities in other ways as well. We know that plants form the base of food chains, since they get energy from the sun. Any food energy that we, or any animal, eats likely got its start as light energy captured by plants. Because plants are the first links in the food chain, we could make predictions about how many organisms can live in different areas; we could call this **biomass**, which is basically like adding up the weight of every living thing in an area. The biomass of your home would be the sum weight of each member of your family, plus any pets and house plants. (Feel free to go find your home's biomass right now!) Comparing a desert and a rainforest, which one do you think has more plants growing in it? If you think rainforests, you'd be right, because they have more water to support plant growth.

Sometimes biologists call this growth of plants **productivity** (or primary productivity, since it's the first step in a food chain). Productivity in an ordinary sense means getting a lot done. If one day you go to school, do you homework, mow the lawn, feed the dog, and clean your room, you were productive. If you wake up and watch TV all day, you are not productive. This is a similar meaning in regards to plants in that if plants grow a lot, they are productive.

So we would predict rainforests have higher productivity, which could support more herbivores, which could then support more carnivores eating those herbivores. Adding up all those extra plants and animals would result in a higher biomass. More nutrients, more rainfall, and better growing conditions all increase the productivity of a community.

Another fascinating factor that influences productivity is the diversity of the system, meaning the number of plant species present. Experiments show that more plant species results in higher productivity (importantly, not more individual plants, *but more types of plants*). So healthy communities also have higher productivity and can support more animal life overall. This means it is important

Rich plant diversity can support a larger variety of animal life. This moose thrives deep in the forests of Sweden.

Invasive Species

An **invasive species** is any animal or plant that is not native to an area that has a destructive effect. Invasive plants might outcompete other plants because they have no natural predators or diseases in the regions they have been introduced in to.

KUDZU

to limit disturbances, like pollution or invasive species, that could reduce plant diversity in an area. Interestingly, succession tends to increase diversity and stability over time. God has put into place a system that can help regulate itself, giving it the balance and order it needs to live and thrive. Amazing!

In our next chapter, we will continue to look at how plants affect life on earth, but more specifically how they influence our own lives.

FOUNDATIONS REVIEW

- ✓ Succession describes changes in a community over time, but in ways that we can predict based on the features of different plants and their interactions with each other. Succession can happen slowly (primary succession), as when a new island is formed in the ocean and abundant plant life takes thousands of years to develop, or rather quickly (secondary succession), as when a forest might regrow in a number of years following a wildfire.

- ✓ Plants are found all over the world in habitats that can be very different, such as dry deserts or wet rainforests. Wherever they grow, they can have an effect on what animals are found in that region. These distinct sets of types of plants and animals that are found in similar habitat types are called biomes.

- ✓ Plant productivity—how much plant growth there is in a region—has a direct influence on the biomass of that region, which is essentially like adding up the weight of all the plants and animals from that region. For example, a rain forest would have higher productivity rates and a higher biomass than a desert because it receives more rain, and thus, has more plants growing more rapidly, attracting more animal species to it than would the desert.

which comes from the Latin word for "change." It refers to change, or growth and formation, in wisdom, faith, and holiness.

Ecological succession is a slow change over time, with a progression toward a more stable, fuller ecological community. We each are called to change and transform ourselves over time to better reflect the likeness of Christ. Part of how we can do this is through learning, which is especially true for all you students reading this right now. You have many years of study ahead (what a great opportunity!), and you can use the knowledge you gain in ways that build up your relationship with God, your family, and your community.

Changing an ecosystem over time isn't just a one-time process; there may be interruptions along the way, such as invasive species, or periods of drought and slow growth, among others. Or progress may even be reset by wildfires or other major disturbances. Instead, there is continual growth and renewal. In the same way, we should always be active in prayer and in our pursuit of knowledge. I hope that you each develop a lifelong love of learning that will help you in your pursuit of *Conversatio*.

A traditional agriculture terraced rice field.

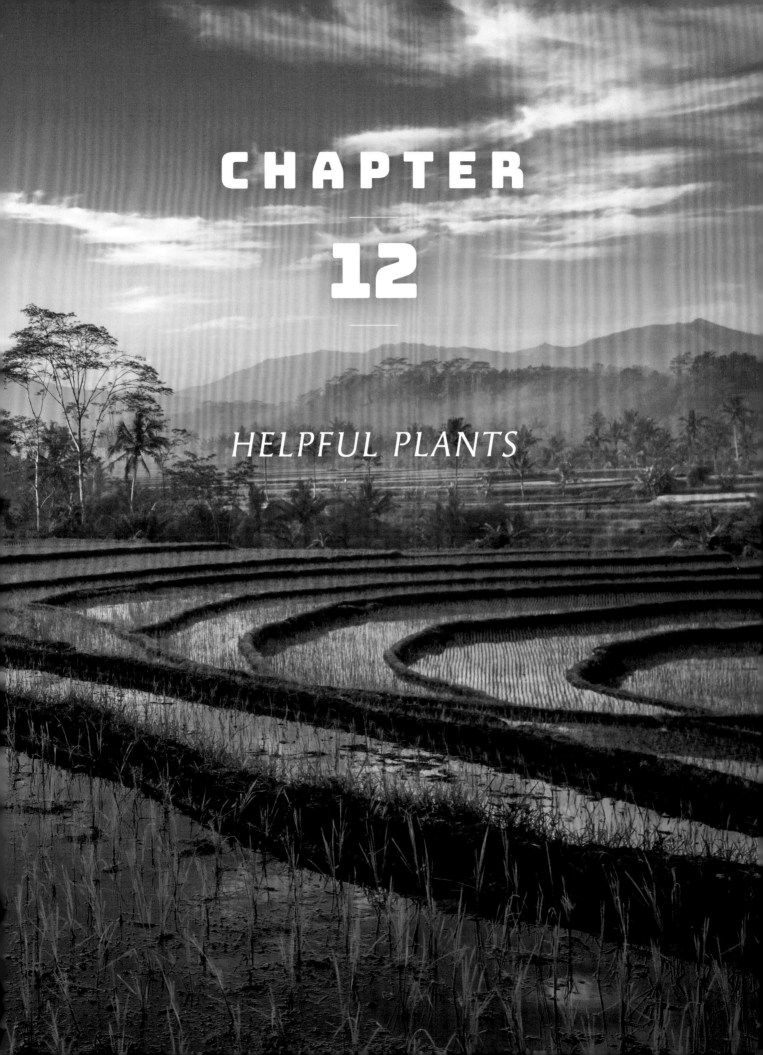

CHAPTER 12

HELPFUL PLANTS

THANK A PLANT!

Did you breathe in today? If yes, thank a plant! Plants release oxygen during photosynthesis, which we and other animals need to breathe. This alone makes plants critical to life on earth, including humans. But we already mentioned this example, and it certainly isn't the only way we rely on plants. A whole textbook could be written about our interactions with plants, but in this chapter, we will try to summarize some important ways that we benefit from plants. Specifically, we will mention four common uses for plants: food, fiber, fuel, and medicine.

FOOD

We already talked a little bit about plants as food in chapter 9, but let's go into more detail here.

Humans started cultivating plants for food around ten thousand years ago, and today more than seven thousand species of plants are grown for human consumption worldwide. We have special relationships with these plants, and those relationships have changed over time through both artificial and natural selection. Here, *artificial* means not occurring naturally, but this does not mean that the food we grow is fake or not natural, but rather that we control the growth and success of certain plants and crops. Instead of being successful by outcompeting other plants, crops are successful when they have traits that we value in foods.

By choosing plants we like best, like the ones that grow fastest or have the biggest fruits, we make it so that only those plants successfully pass on their traits to the next generation. (We make sure to plant their seeds next year.) In this way, corn changed quite dramatically over the centuries from its ancient ancestor, known as **teosinte**.

Plant Fun Fact: The earth has more than 80,000 species of edible plants!

Plants are also used *indirectly* for humans to eat. Plant products are used in animal feed to raise cows, pigs, sheep, chickens, and other animals. As an example, we can look at the cereal grains we learned about from the grass family. About 2.3 billion tons of grain are produced each year: we eat almost 1 billion tons, but another 750 million of those tons are used to feed animals, animals that we then eat (or drink their milk, eat their eggs, etc.). We need a lot of space to grow all of this food, for ourselves and for the animals we keep. Croplands cover more than 10 percent of the earth's ice-free land surface (meaning we don't count Antarctica or parts of the arctic circle in that calculation); this is more than 1.5 billion hectares. Even though 10 percent is a lot of land, knowing the science of how plants grow and produce fruit has helped us produce *more* food on *less* area. Further research and knowledge could help us grow even more food from that same amount of space.

Tea plantations (above) in Munnar, Kerala, South India.

FIBERS

Food is not the only use that we can get from plants. We also use plants or plant products to produce fibers. This does not refer to the dietary fiber listed in the nutritional facts of food (although that is good and we can get fiber from things like broccoli); rather, this refers to the fibers found in things like fabric, rope, and paper products. A cotton t-shirt is made from the fibers of a cotton boll, which is a soft, fluffy casing around the seeds of the cotton plant. Ropes are made from the fiber in hemp or other plants. Both of these plant products are high in cellulose, that structural part of a plant's cells, which helps ropes and fabrics last for a long time. Perhaps this is why these particular plant parts have been used by humans to make fibers.

Another use of plants for fiber is the production of paper. Some of the earliest paper products were made from the papyrus plant. Most paper products now are made from trees, and the average American uses around seven trees per year in paper, wood, and other products. For this reason, it is good to utilize fast-growing plants to produce paper products. Some trees grow faster than others and so would make a good choice, but we have also started utilizing other fiber crops for paper. Bamboo, in particular, grows quickly and uses up less water to grow. Paper can also be made from recycled paper, which can help reduce the number of trees or other plants we use. This means recycling paper is a small way to give back to the plants that help us out.

FUEL

Biofuels are another way that humans use plants. A **biofuel** is any fuel made from biomass (plant or animal material). The use of biofuels is a possible alternative to fossil fuels like oil. Fossil fuels get their name because they come from long dead and buried organisms. If plants are buried for a long time—say, millions of years—they are eventually transformed by the heat and pressure into oil (plankton, on the other hand, would become natural gas). It's fascinating to think that both fossil fuels and biofuels are ways to rely on plants for fuel.

However, fossil fuels take much longer to make. For that reason, they are described as a non-renewable resource. It takes too long to regenerate oil, so once it's used up, we cannot get more. Biofuels are renewable because we can grow more living plants. Biofuels also typically release fewer harmful gasses into the atmosphere, and so are also better for the environment. Corn is one plant that is currently used for biofuel; it can be turned into ethanol for running cars. Who would've thought corn could be used to run your parents' car? Research continues for how to turn other plants into usable fuel.

Plant Fun Fact: Only 1 percent of rainforest plants have been studied for medicinal potential!

MEDICINE

In chapter 10, we learned about how plants had chemical defenses. It turns out that some of the chemicals that plants use for defense or other purposes also have medical applications. Evidence from archaeological digs suggests that humans have used plants in this way for tens of thousands of years. Aspirin, a quite commonly used drug for aches, pains, and fever reduction, is found in the bark of willow trees. The aspirin your parents buy at the store, though, is not likely from a tree. Once we have identified the chemicals (like aspirin) that act as medicine, scientists often figure out a way to make the chemical directly rather than extracting it from plants.

There are many other examples of medicine derived from plants. Menthol, used for sore throats, comes from mint leaves, while quinine is used as a treatment for malaria, coming from the bark of a plant called the chinchona tree. Overall, it is estimated that 40 percent or more of the medicines used today were originally identified in plants!

OTHER HELPFUL USES OF PLANTS

Of course, there are also many more uses for plants: wood for building homes and furniture, beautiful plants used for landscaping, or even simple things like providing shady spots on hot summer days.

Medicine for Animals

Interestingly, animals also use plants for medicinal purposes. Some people think dogs eat grass as a way to ease stomach aches. Chimpanzees, meanwhile, eat certain leaves which have no nutrition but do help kill parasites in their intestines. Orangutans also chew up a particular type of leaf and spread the paste onto themselves. Scientists have shown that this paste has the potential to help with aches and pains, so it's possible the orangutans are using it to treat muscle aches.

But other important services that plants provide for us are a little less obvious. For example, did you know that wetland plants are good at filtering out toxins before they can enter our streams, rivers, and ponds? Certain moss species have also been shown to be successful at absorbing harmful chemicals like lead or arsenic. And in addition to acting as a water filter, plants affect the water cycle on earth in other ways. Water is released into the atmosphere by plants through **transpiration** (water loss out of their stomata). This is why it would be more humid in the middle of a corn field in July than in a parking lot. (An acre of corn can release thousands of gallons per day!) This also means that a major loss in forest or other plant species could result in drier, drought-like conditions in that area.

And yet there are many more important jobs that plants do for us. This includes regulating different parts of our environment. Plants prevent soil erosion so that too much dirt and soil doesn't run off into lakes. Flooding is also reduced by plants, because plants can help slow the movement of water. Diverse communities of plants also support a diversity of other organisms, which ultimately provides a more stable ecosystem.

Plant Fun Fact: The first certified botanical garden was founded by Pope Nicholas III in the Vatican City in AD 1278!

Overall, any of these things that plants do for us are called **ecosystem services**. Different estimates have been put forward for how valuable these services are to humans (trillions of dollars per year!), but regardless of the dollar amount, it's clear that plants are a critical part of our lives.

Finally, another set of ecosystem services that plants, and other organisms, provide for us is a category called **cultural services**. It may be harder to put a precise value on these, which include the value we place in having beautiful, natural areas to explore. Natural areas can give us great opportunities to learn about God by exploring his works. Hiking, camping, and exploring can also be a great way to build family memories and friendships and benefit the community. We know God is omnipresent, so of course we can talk with him through prayer wherever we go, but finding a quiet place in nature can be another great way to reflect and pray. A few of my favorite places for prayer are adoration chapels I have found that have large windows opening to a forest or other natural area. As a biologist, I love nature, but how much greater is it to sit in these natural areas together with the real presence of Christ in the Eucharist!

As you can see, plants are vital in so many ways to the survival of humanity. We depend on them for our air, our food, our medicine, and so many other important things that affect our daily lives.

Before wrapping up this unit on plants, make sure to flip ahead and read the conclusion.

FOUNDATIONS REVIEW

✓ In addition to helping filter the air that we breathe, plants are incredibly important to life on earth, including for human beings. Four main categories they aid us include food, fiber, fuel, and medicine.

✓ Plants are an abundant source of food for us but also for the animals that we depend on for food. Additionally, plant fibers are used to make things like rope, thread, and paper, and certain plants can be used for biofuels to help run our cars. Finally, throughout history plants have widely been used for medicinal purposes; overall, it is estimated that 40 percent or more of the medicines used today were originally identified in plants.

✓ Even beyond these four main categories, plants serve many other purposes, including the wood we use to build homes and furniture, filtering out toxins in bodies of water across the globe, and for our own personal peace and sense of happiness in the form of nature hikes and city parks, among others.

Conclusion

Congratulations! You've made it to the end of our book on plants! We discussed a lot of different topics, including the key features plants share, the differences between plant groups that help us categorize them, and the importance of plants to life on earth.

I hope that you enjoyed our tour of the plant kingdom. Some people may think animals are more interesting since they move around and do stuff, but now you can tell them that plants are doing all sorts of cool things too! They fight for resources, defend themselves against animals, take care of their young, and more. I am always filled with wonder when I learn more about the world God created.

Completing this unit on plants is a great start to being a biologist, or plant ecologist, learning about the world God has given to us. But now you must use what you learned here and go explore nature. Get up close and personal with a plant! What do you see when you look closely at its parts? Are there any insects using it for food or a place to live? We saw how important plants are to our world, so scientists like you and me should do what we can to understand these wonders of nature!

TROPICAL LEAVES

AMAZING PLANT FACTS

- Unlike animals, who mostly go out and find their food, plants generate their own food interiorly through a process called photosynthesis, which uses the sun's energy to create food sugars that the plant can "eat."
- Through photosynthesis, plants sustain the entire food chain, since animals (including humans) eat plants, but also eat animals who eat plants. All energy in the food chain enters through the process of photosynthesis.
- Through the process of photosynthesis, plants release oxygen into the air, which we then breathe. Without plants, we would not be able to live!
- There are around 375,000 species of plants described on earth, and likely even more we haven't discovered yet!
- Duckweed, a pond-living plant, is one of the smallest plants in the world, at only a fraction of an inch in size. In contrast, the sequoia trees (or redwoods) can be more than three hundred feet tall!

Plant Fun Fact:
Oak trees are struck by lightning more than any other tree!

- The most well-known tree in the world is named General Sherman, found in the state of California. Its trunk is 103 feet around and it is over 50,000 cubic feet in volume; that means, by very rough approximation, it would take a fleet of 400 to 500 minivans to haul it away if it fell in the forest, with all the seats folded down!
- The world food supply relies on 150 or more plant species; without these plants turning light energy into their own food, we wouldn't have any food.
- About half of the energy we humans get from food comes from three major crops: rice, wheat, and corn.

- The first cultivation of plants for food may have happened over twenty thousand years ago! We didn't develop agriculture as we know it today, with large fields and farms, until much later, but still thousands of years ago.

- In a forest in Utah, there is a large grove of more than forty-five thousand aspen trees. This may not be all that remarkable, until you realize that each tree is genetically identical. It is thought that each of the trees has grown outward as a clone from a single individual. The trees even have connections to each other underground. So rather than a single tree, it is more like many trees linked together to form one individual, or one "super-organism." This "individual" even has its own nickname: Pando. Together, these trees weigh more than 6,500 tons!

- A pigment called chlorophyll gives plants their green color. When light hits a plant's chloroplasts (where the chlorophyll is contained), they absorb most of the light but reflect the green light, which our eyes detect as a green color.

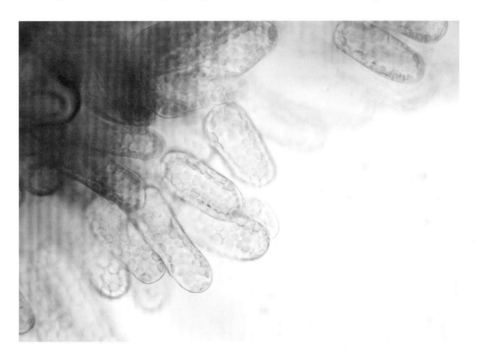

Remember:
Not all plants are green, as some rare ones don't have many chloroplasts.

- The group of plants called bryophytes is unique because they have no seeds, no roots, no flowers, and no real leaves. An example of a common bryophyte is moss.

- Some liverwort species (bryophytes) can regrow from a small piece; you could break a liverwort into two pieces and end up with two plants!

- Most plant species on earth are vascular plants; eight times as many, in fact, with nearly 400,000 species, and more being discovered all the time.

- In the Kalahari Desert, there is a plant called the Shepherd plant whose roots can go over two hundred feet deep into the ground!

- The orange part of the carrot that we eat is a taproot (central root) of the carrot plant.

Plant Fun Fact:
We think of carrots as orange, but some can be purple!

- Maple trees store things in their roots in the winter (sugar, water, nutrients) and send those things up into the top of the tree through the vascular system; this provides everything needed to make leaves in the spring. We take advantage of this by harvesting some of the sap they move upwards to make maple syrup!
- The horsetail plant, given its name because of its resemblance to a horse's tail, have spores that can walk! The spores have four little spindly "legs" (called elaters) that curl and uncurl with changes in moisture, and the movement of the legs helps the spores move around (and even sometimes jump!).
- The largest fern species can grow over thirty feet tall!
- Cactuses have spikes, which are essentially modified leaves that protect them from hungry animals.
- Have you ever wondered how plants know to grow upwards? Believe it or not, scientists think that plants have cells that work as gravity detectors! Imagine a jar filled with water and rocks. If you turn the jar over, the rocks will sink down to the new "bottom" of the jar. These plant cells are similar in that they have small "stones" that always sink to the bottom, telling the plant where to send its leaves, away from the "stones."
- Seeds need to travel to new places so plants can grow in new areas. How do they move? Some hitch a ride on animals—you may have encountered burrs that get stuck to your pants or a pet's fur. These are seeds that rely on mammals to get them to someplace new. Other plants rely on animals, but by trading something for the ride. Certain fruits that contain seeds are eaten by animals (free lunch for the animal!) but the seeds are then passed out the other end of the animal. Animal feces also makes good fertilizer (lots of nitrogen), so that's just an added bonus for the young plant. Another way to move around is by floating in the wind. Dandelion seeds have little parachutes to catch the wind and spread out the seeds.

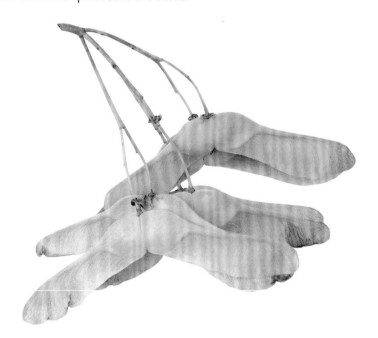

- Conifer trees are known to live a long time. A bristlecone pine tree in eastern California, nicknamed Methuselah, is almost five thousand years old!
- Flowering plants represent more than 80 percent of known and described plant species.
- Some orchids may trick bees into coming to them by looking like a place to take shelter, or even by resembling another bee. There are likely over one thousand species of orchids that trick pollinators into visiting their flowers.
- Examples from the magnoliid group include plants that you may be familiar with. Many members of this group contain chemicals in their tissues that give them particular smells and tastes; thus, humans have used them for cooking all over the world, including black pepper, cinnamon, and nutmeg.
- More than a third of the earth's total land area is covered with grass species!
- One cool set of species from the eudicot group is the acacias. Some acacia plants have a mutualism with ants. The ants protect the plant by biting animals that want to eat the plant, and, in return, the plant provides nectar to the ants. In a weird twist, though, scientists discovered that the ants actually cannot leave the plant once they drink the nectar; they become totally reliant on that plant for food.
- There are some plants that are carnivorous, meaning they eat meat. This might not sound like what you think, though. They simply eat insects by luring them over with the promise of something they want (perhaps sweet nectar), but then they drown them in pools of liquid or poison them with a chemical substance, and then digest them. Because this counts as animal tissue, we say they are carnivorous.

- Some giant kelp can grow one hundred feet up from the ocean floor!
- A particular pitcher plant species, *Nepenthes hemsleyana*, has a mutualism with bats. The plant has specialized leaf structures to help echolocating bats

find the plants. In turn, the bats use these pitchers as a daytime roost. While they hang upside down in the plant to sleep, they also poop, which the plant digests as a source of nutrients. Several other pitcher plant species offer sugary nectar to small mammals who visit the plant to lick up the nectar. The interesting thing is that the nectar is positioned so that the animal is sitting right over the top of the pitcher while it eats, so that it's like a tiny toilet for shrews (a mole-like animal) that offers them a snack!

Remember:

Cucumbers, tomatoes, and green peppers all develop from flowers and have seeds, which makes them a fruit!

- In botany, a fruit has a specific definition: anything that contains seeds and develops from the ovary of the flower. This means that many things we call vegetables are actually fruits—cucumbers, tomatoes, and green peppers all develop from flowers and have seeds. Perhaps more surprisingly, eggplant, olives, pumpkins, and even wheat (and other grains) are all fruits.

- A strawberry is not actually a berry, nor is it even a fruit!

- Tapirs are known to deposit seeds through their dung in poor forest habitats, like places where a fire has come through. In this way, they can actually help rainforests recover from a fire or some other situation where it was destroyed, and rainforests are an important aspect of sustaining life on earth, including for us, since they help provide us with the oxygen we need to breathe.

- Horseradish plants have two chemicals that, if kept separate, don't really pose any danger to organisms. But when an herbivore (including a human) chews up the plant, it damages the cells. Little containers that hold the chemicals burst, mixing the chemicals together. A chemical reaction then follows, and a new chemical (allyl isothiocyanate) is formed; this irritates eyes and lungs and tastes terrible (a small amount of this chemical gives horseradish sauce its distinct kick).

- Peppers have different amounts of a chemical called capsaicin that causes a sensation of pain when it binds to some of our taste receptors (jalapeños have some, ghost peppers have a lot). It turns out, though, that only mammals experience pain from capsaicin, because we have the right receptors. Birds, on the other hand, do not have receptors for the capsaicin to bind to, and so they can eat spicy peppers with no issues.

- Roses technically have prickles, not thorns.

- Plants compete with neighbors for sunlight and other resources. Some actually release chemicals that prevent other plants from growing nearby. If no competition can grow, all water and nutrients in the area are available. Black walnut trees are a good example of this form of chemical warfare. Parts of the black walnut tree release chemicals, which prevents plants from growing under and around the tree.

- Humans started cultivating plants for food around ten thousand years ago, and today more than seven thousand species of plants are grown for human consumption worldwide.

- About 2.3 billion tons of grain are produced each year: we eat almost 1 billion tons, but another 750 million of those tons are used to feed animals, animals that we then eat (or drink their milk, eat their eggs, etc.). We need a lot of space to grow all of this food, for ourselves and for the animals we keep. Croplands cover more than 10 percent of the earth's ice-free land surface; this is more than 1.5 billion hectares. Even though 10 percent is a lot of land, knowing the science of how plants grow and produce fruit has helped us produce *more* food on *less* area.

- Plants are used in the production of rope, thread, and paper. Some of the earliest paper products were made from the papyrus plant. Most paper products now are made from trees, and the average American uses around seven trees per year in paper, wood, and other products.

- Plants are used in the production of biofuels. Corn is one plant that is currently used, being turned into ethanol for running cars.

- Aspirin, a quite commonly used drug for aches, pains, and fever reduction, is found in the bark of willow trees. The aspirin your parents buy at the store, though, is not likely from a tree. Once we have identified the chemicals (like aspirin) that act as medicine, scientists often figure out a way to make the chemical directly rather than extracting it from plants. Menthol, used for sore throats, comes from mint leaves, while quinine is used as a treatment for malaria, coming from the bark of a plant called the chinchona tree. Overall, it is estimated that 40 percent or more of the medicines used today were originally identified in plants!

- Wetland plants are good at filtering out toxins before they can enter our streams, rivers, and ponds, and certain moss species have also been shown to be successful at absorbing harmful chemicals like lead or arsenic.

- Plants help balance the water cycle on earth. Water is released into the atmosphere by plants through transpiration (water loss out of their stomata). This is why it would be more humid in the middle of a corn field in July than in a parking lot. (An acre of corn can release thousands of gallons per day!) This also means that a major loss in forest or other plant species could result in drier, drought-like conditions in that area.

KEY TERMS

Alternation of Generations – *Chapter 3*: The phenomenon by which plants have two "forms" (or types) that each reproduce to develop into the other form.

Angiosperms – *Chapter 6*: Flowering plants; represent more than 80 percent of known plant species.

Remember:
Flowers are not just pretty to look at; they also serve valuable roles for the plant.

Annual plant – *Chapter 7*: The opposite of a perennial plant; plants that release seeds and then die, but the seeds survive the winter and then grow in the spring.

Anther – *Chapter 6*: The part of the stamen in a flower that produces pollen.

Anthropomorphism – *Chapter 9*: Using language where we talk about animals or plants almost with human like qualities.

Apical meristems – *Chapter 3*: A common feature in land plants where the cells divide and grow at the root and the shoot tips (the bottom and top of the plant).

Autotrophs – *Chapter 1*: An organism that can generate its own food, like plants.

Berry – *Chapter 9*: A simple fruit that develops from a single flower with a single ovary, and that has multiple seeds.

Biofuel – *Chapter 12*: Any fuel made from biomass (plant or animal material).

Biomass – *Chapter 11*: The added sum weight of all living things in a given area or habitat.

Biomes – *Chapter 10*: Distinct sets of types of plants and animals that are found in similar habitat types (desert, rainforest, etc.)

Botany – *Chapter 9*: The study of plants.
Bryophytes – *Chapter 3*: Small, flowerless, seedless plants, with no real roots or leaves or vasculature. A common example is moss.

Capsaicin – *Chapter 10:* A chemical found in some plants (like peppers) that causes a sensation of pain when it binds to some of the taste receptors of mammals.
Carnivorous plants – *Chapter 8*: Plants that eat insects, which is considered animal tissue, so these are plants that eat meat.
Cellulose – *Chapter 1 & 8*: A tough material or substance that makes up the cell walls of plants; helps them remain stiff and upright.
Cell wall – *Chapter 10*: The outer part of a cell that contains cellulose and other structural components that help protect the cell.
Chitin – *Chapter 1 & 8*: A material or substance that makes up the cell walls of fungal cells.
Chlorophyll – *Chapter 2*: A pigment found within the chloroplasts of plants that helps with the process of photosynthesis by absorbing sunlight and capturing the energy within it; chlorophyll is what gives plants their green color; when light hits these chloroplasts, they absorb most of the light but reflect the green light, which our eyes detect as a green color.
Chloroplasts – *Chapter 2*: Tiny structures—organelles—in plant cells where photosynthesis is conducted; they house small molecules called pigments, most importantly the pigment of chlorophyll, which absorbs the sunlight and captures the light energy.
Community – *Chapter 11*: When used in biology, all the organisms in an area that interact together.
Cotyledon – *Chapter 7*: The first tiny leaf of a baby plant.
Cultural services – *Chapter 12*: Any way plants benefit human culture, such as gardens and city parks.
Deciduous trees – *Chapter 4*: Trees that drop their leaves and grow new ones each year.

Ecosystem services – *Chapter 12*: The ways in which plants benefit ecosystems, such as filtering toxins out of bodies of water.

Endocarp – *Chapter 9*: The very inside, or core, of a fruit, which provides protection for the seeds and serves to help the parent plant tissue (which the fruit develops from) and the offspring (the seeds).

Endosperm – *Chapter 5*: A tissue found inside the seed that provides the developing plant with nutrients.

Energy – *Chapter 2*: The force that gives living things the ability to "work," or to function and stay alive; can come in different forms, such as light energy (from the sun) or chemical energy, among others.

Exocarp – *Chapter 9*: The outer layer of a fruit, like the skin or peeling of an apple or banana; provides protection for the fruit and seeds.

Fibrous roots – *Chapter 4*: A branching, tangled web of roots (different from tap root).

Filament – *Chapter 6*: The slender stalk part of a flower, found in the stamen, that supports the anther, where pollen is produced.

Food chain – *Chapter 1*: The tiered series of organisms that are all dependent on the next as a source of food (animals eat plants, we eat animals, etc.).

Fruit – *Chapter 9*: Anything that contains seeds and develops from the ovary of the flower.

Funiculus – *Chapter 9*: In flowering plants, the stalk that attaches the ovule to the placenta; acts like a kind of umbilical cord.

Germination – *Chapter 5*: The process of a seed sprouting and starting to grow.

Gymnosperms – *Chapter 5*: A type of plant that has vasculature and seeds.

Heterotrophs – *Chapter 1*: An organism that cannot produce its own food; this includes animals.

Invasive species – *Chapter 11*: Any animal or plant that is not native to an area that has a destructive effect.

Kingdom Plantae – *Chapter 8*: Another name for the plant kingdom.

Lichen – *Chapter 10*: A crusty or sometimes leaf-like, flat growing organism found on bare rocks and trees; it can grow practically anywhere, and is a partnership between fungus and algae.

Mesocarp – *Chapter 9*: The fleshy middle of the fruit, between the skin and core of the apple; usually offers the most nutrition to humans or animals that might want to eat the fruit.

Multicellular – *Chapter 1*: When an organism has more than one cell; one of the primary traits of plants.

Multicellular, dependent embryos – *Chapter 3*: The phenomenon whereby plants of the next generation get their start while still within their parent plant.

Mutualism – *Chapter 2 & 6*: When two organisms work together and each benefit from the partnership.

Nectar – *Chapter 6*: A sweet, sugary substance found in some flowers that helps to attract pollinators.

Nectar guides – *Chapter 6*: Patterns that help lead the animal pollinators (like a foraging bee) to the center of the flower.

Non-photosynthetic plants – *Chapter 8*: Plants that do not conduct photosynthesis.

Non-vascular plants – *Chapter 3*: Plants that do not have vasculature, like bryophytes.

Ovary – *Chapter 6*: Female organ of a flower, found in the pistil.

Parasite – *Chapter 8*: Any organism that feeds off another organism.

Perennial plants – *Chapter 7*: Plants that live multiple years (the opposite of an annual plant); to do this, they have storage organs (called bulbs or corms) to store things through the winter before regrowing in the spring.

Petal – *Chapter 6*: Modified leaves in angiosperms found on the flower; often are brightly colored and are meant to attract pollinators; contain the stamen and pistil.

Phenology – *Chapter 11*: The seasonal timing of events like flowering, when fruits set, and other natural events.

Phloem – *Chapter 4*: Vascular tissue in plants that helps move things from the top of the plant down.

Pistil – *Chapter 6*: The female reproductive part of the flower found in the petals and bearing a vase-like shape; consists of the stigma, style, and ovary.

Remember:
Roses do not actually have thorns, as most people commonly think, but prickles.

Photosynthesis – *Chapter 1 & 2*: The process by which plants use the sun's energy to generate their own food.

Pollination – *Chapter 6*: The process of fertilization, or reproduction, in gymnosperm and angiosperm plants.

Pollinators – *Chapter 6*: Animals that help pollinate flowers by carrying pollen grains from one flower to the next.

Prickles – *Chapter 10*: Tiny spikes that form from the outer surface of the plant tissue on certain plants that gives them a structural defense against animals who would eat the plant.

Productivity – *Chapter 11*: In regards to plants, the rate at which a plant or group of plants grows.

Receptacle – *Chapter 9*: The part of the flower where the sepals and petals attach.

Rhizoids – *Chapter 3*: Small hair-like structures in plants like bryophytes that help anchor them to the ground.

Rhizomes – *Chapter 9*: Stems that grow outward away from the plant underground. Some rhizomes we use in cooking include ginger, turmeric, and arrowroot.

Roots – *Chapter 4*: In vascular plants, the organs that anchor the plant into the ground and help it soak up water and nutrients from the surrounding soil.

Seed – *Chapter 5*: Essentially, a baby plant; contain the plant embryo that represents the next generation as it grows into an adult.

Seed coat – *Chapter 5*: The protective outer layer of a seed.

Sepals – *Chapter 6*: The outermost portion of a flower; provide protection from cold weather or other things that could injure the flower.

Spines – *Chapter 10*: Modified leaves that appear like spikes on certain plants, like the cactus, that offer it protection.

Stamen – *Chapter 6*: The male reproductive organ of the flower that produces pollen and is found within the petals; consists of the filament and anther.

Stigma – *Chapter 6*: The top most part of the pistil in a flower where pollen germinates.

Stoma (plural **stomata**) – *Chapter 2*: Small openings in the stem or leaves of plants to help them exchange gasses with the atmosphere.

Style – *Chapter 6*: The neck of the "vase" part of the pistil of a flower connecting the stigma and ovary.

Sori – *Chapter 4*: Yellow structures on ferns that hold spores, found on the underside of leaves.

Spores – *Chapter 4*: The reproductive cell found in plants that do not have seeds.

Succession – *Chapter 10*: The change in a plant community over time, generally toward a more diverse and complex community. Can occur through **primary succession** (slower, starting from no plant life) or **secondary succession** (faster, regrowth after disturbance).

Taproot – *Chapter 4*: A central root found in some plants.

Teosinte – *Chapter 12*: The ancient ancestor of corn.

Thorn – *Chapter 10*: Spikes found on the stem and branches of some plants to offer protection from animals who want to eat it.

Trait – *Chapter 1*: A distinguishing characteristic or physical quality; in this case, in regards to plants.

Transpiration – *Chapter 12*: The process of water being released into the atmosphere by plants through their stomata.

Tuber – *Chapter 9*: A swelling of plant tissue that is used to store water, starches, sugars or other things the plant needs.

Unicellular – *Chapter 1*: When an organism has only one cell.

Vascular plants – *Chapter 4*: Plants that have vasculature.

Vasculature – *Chapter 3 & 4*: A transport system in plants, a series of tubes, that moves water and other nutrients all throughout the plant.

Xylem – *Chapter 4*: Vascular tissue found in plants that moves water and minerals and nutrients from the root up through the plant.

IMAGE CREDITS

Front cover Maple tree seeds, Gavran333 / Two winged maple seeds, joanna wnuk / Lemon with leaves, Noam Armonn / sycamore seeds, majaan / ficus branch, rubber plant, ANCH / autumn maple tree, Potapov Alexander / Purple orchid flower, Alexander Denisenko / Reaching hands from The Creation of Adam of Michelangelo, Freeda Michaux © Shutterstock.com

pI Young green plant © RachenArt, Shutterstock.com
pII Miniature succulent plants © dinodentist, Shutterstock.com
pV Space background with many stars © Sundays Photography, Shutterstock.com
pVI Tracktor plowing field © Valentin Valkov, Shutterstock.com
pVIII Peripatus (Velvet Worm) © Dr Morley Read, Shutterstock.com
pIX Homalomena Wallisii © ittisak boonphardpai, Shutterstock.com
pX Beautiful alpine meadow with a variety of wild flowers © Artyart, Shutterstock.com
pXII-1 Summer flowers at local farm market © David Kay, Shutterstock.com
p2 Corn © lovelyday12, Shutterstock.com
p2 Shrubs and green lawns © Yarygin, Shutterstock.com
p2 Oak tree © rsooll, Shutterstock.com
p2 Pink waterlily or lotus flower in pond © Praew stock, Shutterstock.com
p3 Green leaf glowing in the sunlight © Chyrko Olena, Shutterstock.com
p4 Fresh vegetables, fruits © Lotus_studio, Shutterstock.com
p5 Hiker in Sequoia national park in California © My Good Images, Shutterstock.com
p6 Picacho Peak at sunset, surrounded by blooming desert © Anton Foltin, Shutterstock.com
p6 Obscure morning glory (Ipomoea obscura) © Chansom Pantip, Shutterstock.com
p7 Small seed on a finger © Zapylaiev Kostiantyn, Shutterstock.com
p8-9 Rays of light falling through green canopy of a beech tree © Smileus, Shutterstock.com
p10 Fall photo of world's oldest organism, a grove of Populus tremuloides (Quaking Aspen) sharing one root system, from Fish Lake National Forest website / J Zapell / [public domain], via Wikimedia Commons
p11 Process of Tree Produce Oxygen illustration © BlueRingMedia, Shutterstock.com
p12 Chloroplast, plant cell organelle © Aldona Griskeviciene, Shutterstock.com
p13 Elysia chlorotica, Eastern Emerald Elysia. Wallace Creek, Honga River, Dorchester County, MD - 07/12/17. Photo by Robert Aguilar, Smithsonian Environmental Research Center. / Elysia chlorotica_0176 / Smithsonian Environmental Research Center, Creative Commons (CC), via Wikimedia Commons
p14 Tall tree © Lev Kropotov, Shutterstock.com
p15 Elemental oxygen (O2), molecular model © StudioMolekuul, Shutterstock.com
p15 Basil © grey_and, Shutterstock.com
p16-17 Green leaves and sunlight © vovan, Shutterstock.com
p18-19 Waterfall and mossy rocks © rdonar, Shutterstock.com
p20 Mangrove trees along the sea © happystock, Shutterstock.com
p20 Arizona Desert Cactus of Opuntia Genus © You Touch Pix of EuToch, Shutterstock.com
p20-21 Stages of growth and flowering orange daylily © Sarycheva Olesia, Shutterstock.com
p22 Green ferns in forest © sandart, Shutterstock.com
p23 Natural moss © Tatiana Maksimova, Shutterstock.com
p24 Water plants (rigid hornwort) © MDWines, Shutterstock.com
p24 Marchantia polymorpha liverwort gametospores © IanRedding, Shutterstock.com
p24 Spore capsules © Henri Koskinen, Shutterstock.com
p25 The Sower / English School, (19th century) / English / Bridgeman Images
p26-27 Golden section of fern © Yarygin, Shutterstock.com
p29 Big tree roots © siambizkit, Shutterstock.com
p29 Green oak leaf © Fedorov Oleksiy, Shutterstock.com
p29 Branch of Pinus bungeana © Lukas Gojda, Shutterstock.com

IMAGE CREDITS

p30 Autumn leaves © Galaxy love design, Shutterstock.com

p31 Horsetail © aga7ta, Shutterstock.com

p31 Giant Saguaro in Sonoran Desert © Nate Hovee, Shutterstock.com

p32 Hand holding a young green plant © Iryna Rasko, Shutterstock.com

p33 Saints Cosmas and Damian, miniature from the Grandes Heures of Anne of Brittany / circa 1503-1508 / Jean Bourdichon (1457–1521) / Bibliothèque nationale de France / [public domain], via Wikimedia Commons

p34-35 Close up of a dandilion © PS Media House, Shutterstock.com

p36 Cone and seeds © nulinukas, Shutterstock.com

p37 Anatomy of a bean seed © Designua, Shutterstock.com

p38 Dandelion seeds blowing in the wind © solarseven, Shutterstock.com

p39 Pine tree © Volodymyr Goinyk, Shutterstock.com

p39 Cycad tree © Rob Atherton, Shutterstock.com

p39 Methuselah - The oldest living Great Basin bristlecone pine (Pinus longaeva) tree in the world © doliux, Shutterstock.com

p40 Ginkgo biloba green leaves © sunnyfrog, Shutterstock.com

p40 Bud of plant germination © fotoslaz, Shutterstock.com

p40 Pong pong seed © warat42, Shutterstock.com

p40 Small green seedlings © Bignai, Shutterstock.com

p41 Plant Illustrations © Morphart Creation, Shutterstock.com

p41 Annie Chambers Ketchum / 1890 / Gaines, B. O. (1890) domain The B.O. Gaines History of Scott County, B.O. Gaines Printery, p. 304 / G. O. Gaines / [public domain], via Wikimedia Commons

p42-43 Passion fruit flower © Matjoe, Shutterstock.com

p44 Plant reproductive system diagram, annotated © grayjay, Shutterstock.com

p45 Common Eastern Bumble Bee covered in pollen © Elliotte Rusty Harold, Shutterstock.com

p46 Hummingbird, feeding on a flower © Henk Bogaard, Shutterstock.com

p47 Mimulus flower photographed in visible light (left) and ultraviolet light (right) showing a nectar guide visible to bees but not to humans, Plantsurfer, Creative Commons (CC), via Wikimedia Commons

p48 Bogus Yucca Moth - Prodoxus species, Carolina Sandhills National Wildlife Refuge, McBee, South Carolina / Judy Gallagher / https://www.flickr.com/photos/52450054@N04/17052239225 / Creative Commons (CC), via Wikimedia Commons

p48 Yucca plant © Ancha Chiangmai, Shutterstock.com

p48 Purple Bougainvillea © Chansom Pantip, Shutterstock.com

p49 Erin Go Braugh. St. Patrick's Greetings / circa 1909 / Missouri History Museum / http://images.mohistory.org/image/6AA90185-ACFA-C07E-985F-567B0808A839/original.jpg / Gallery: http://collections.mohistory.org/resource/152452 / [public domain], via Wikimedia Commons

p50-51 Field of flowers © Marina Grau, Shutterstock.com

p52 Magnolia close up © FraziG, Shutterstock.com

p53 Bamboo leaf © Boonchuay1970, Shutterstock.com

p53 Mulberry leaf © Lepas, Shutterstock.com

p54 Tulip in a pot © Holiday.Photo.Top, Shutterstock.com

p54 Green Grass © Anan Kaewkhammul, Shutterstock.com

p55 Sunflower © Jit-anong Sae-ung, Shutterstock.com

p55 Green clovers © Elena11, Shutterstock.com

p56 Ant — Pseudomyrmex species, on Bull Thorn Acacia (Acacia cornigera) with Beltian bodies / https://www.flickr.com/photos/52450054@N04/8505045055/ Judy Gallagher / Creative Commons (CC), via Wikimedia Commons

p56 Cinnamon sticks with powder © Valery121283, Shutterstock.com

p57 Gregor Johann Mendel © German Vizulis, Shutterstock.com

p58-59 Lion's Mane Mushroom © Fotografiecor.nl, Shutterstock.com

p60 Black fungus © akepong srichaichana, Shutterstock.com

p61 Giant kelp © Ethan Daniels, Shutterstock.com

p62 Puffball fungus © godi photo, Shutterstock.com
p62 Honey fungus © Zbigniew Dziok, Shutterstock.com
p63 22 image focus stack of a fly being processed by a d. capensis plant, Ben pcc, Creative Commons (CC), via Wikimedia Commons
p63 Carnivore plant Nepenthes in detail © Kuttelvaserova Stuchelova, Shutterstock.com
p64 Monotropa uniflora © Fang ChunKai, Shutterstock.com
p64 Chanterelle mushrooms © Spalnic, Shutterstock.com
p65 Portrait of St. Francis Xavier praying, c.1600 (oil on panel) / Unknown Artist / By permission of the Governors of Stonyhurst College / Bridgeman Images
p66-67 Vegetables and fruits © Brian Chase, Shutterstock.com
p68 Orangutan eating a leaf © Kambiz Pourghanad, Shutterstock.com
p69 White horse grazing in a meadow of green grass © Laurent CHEVALLIER, Shutterstock.com
p69 Beetroot © Anna Kucherova, Shutterstock.com
p70 Berries © Bojsha, Shutterstock.com
p71 Green apple © Anastasiia Skorobogatova, Shutterstock.com
p71 Tapir in a river © Guilherme Battistuzzo, Shutterstock.com
p72 Central American agouti © Rosa Jay, Shutterstock.com
p73 St. Paul Preaching at Athens (cartoon for the Sistine Chapel) (PRE RESTORATION) / Raphael (Raffaello Sanzio of Urbino) (1483-1520) / Italian / Victoria & Albert Museum, London, UK / Bridgeman Images
p74-75 Prickly pear cactus © unjiko, Shutterstock.com
p76 Horseradish sauce in wooden spoon © xpixel, Shutterstock.com
p76 Horseradish © Orest lyzhechka, Shutterstock.com
p77 Jalapeno peppers © Binh Thanh Bui, Shutterstock.com
p77 Caterpillar © kuncron, Shutterstock.com
p78 Honey Locust Thorns © Anna Chudinovskykh, Shutterstock.com
p78 Green Cactus © travelview, Shutterstock.com
p78 Rose stem prickles © Sandeep-Bisht, Shutterstock.com
p79 Golden barrel cactus © Saran_Poroong, Shutterstock.com
p79 The largest black walnut tree in North America, in Sauvie Island, Oregon, Gobywalnut, Creative Commons (CC), via Wikimedia Commons
p80 Opuntia cactus © arka38, Shutterstock.com
p81 Sparrows © Bachkova Natalia, Shutterstock.com
p82-83 Lush rain forest © Simon Bennett, Shutterstock.com
p84 Falling Dominoes © Magcom, Shutterstock.com
p85 Volcano eruption © Fotos593, Shutterstock.com
p85 Lava entering the ocean © Yvonne Baur, Shutterstock.com
p85 Volcanic made Island © ARWood, Shutterstock.com
p85 New leaves burst forth from a burnt tree © CHOKCHAI POOMICHAIYA, Shutterstock.com
p86 Lichen © HAOS, Shutterstock.com
p87 Subtropical forest in nepal © Teo Tarras, Shutterstock.com
p87 Cactus in desert © Ilyshev Dmitry, Shutterstock.com
p87 African savannah landscape © Maciej Czekajewski, Shutterstock.com
p87 Spruce Tree Forest © dugdax, Shutterstock.com
p88 Big male Bull moose © Petr Salinger, Shutterstock.com
p89 kudzu © J.K. York, Shutterstock.com
p89 Island © tawin bunkoed, Shutterstock.com
p90 St. Benedict / Piazzetta, Giovanni Battista (1683-1754) / Italian / San Nicolo del Lido, Venice, Italy / Bridgeman Images
p92-93 Traditional agriculture terraced rice field © aulia medika, Shutterstock.com
p94 A field of corn © Mary Lane, Shutterstock.com
p95 Tea plantation © Marisa Estivill, Shutterstock.com
p95 tack of colorful clothes © New Africa, Shutterstock.com
p96 Algae seaweed research, biofuel industry science © Chokniti Khongchum, Shutterstock.com

IMAGE CREDITS **115**

p96 Herbs and medicinal bottles © Alexander Raths, Shutterstock.com
p97 Wooden house built from logs © Aigars Reinholds, Shutterstock.com
p97 Dog eats grass © Daria Photostock, Shutterstock.com
p98 Wood timber © artnami, Shutterstock.com
p99 Tropical leaves © NABODIN, Shutterstock.com
p100 Duckweed © yut4ta, Shutterstock.com
p100 Sequoia Tree © Stephen Moehle, Shutterstock.com
p101 Water plant under the microscope, chlorophyll © Barbol, Shutterstock.com
p102 Maple wings © Lena Pronne, Shutterstock.com
p103 Venus fly trap © josehidalgo87, Shutterstock.com
p105 Ecological wetland © HelloRF Zcool, Shutterstock.com
p106 Field of flowers © Orientaly, Shutterstock.com
p107 Haircap moss © Anest, Shutterstock.com
p108 Fibrous root system © Ayah Raushan, Shutterstock.com
p108 Taproot system © Sahana M S, Shutterstock.com
p109 Closeup of flower pistil and petal © Tao Jiang, Shutterstock.com
p110 Iris Rhizomes © jelloyd, Shutterstock.com
p111 Fern spores © thecloudysunny, Shutterstock.com
p116 Coconut leaves © Dewin ID, Shutterstock.com
p116 Red-green cactus © Ollga P, Shutterstock.com
Back cover Hand holding an onion, Charoen Krung Photography / Wheat, xpixel / Cotton plant flower, Valentina Razumova © Shutterstock.com

The Foundations of Science introduces children to the wonders of the natural world in light of God's providential care over creation.

Too often we hear that science is in conflict with faith, but Pope St. John Paul II wrote that faith and science "each can draw the other into a wider world, a world in which both can flourish." Foundations seeks to spawn this flourishing in the hearts and minds of young readers, guiding them into a world that will delight their imaginations and inspire awe in the awesome power of God.

This eight-part series covers an extensive scope of scientific studies, from animals and plants, to the galaxies of outer space and the depths of the ocean, to cells and organisms, to the curiosities of chemistry and the marvels of our planet. Still more, it reveals the intricate order found beneath the surface of creation and chronicles many of the Church's contributions to science throughout history.

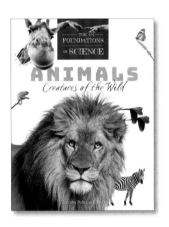

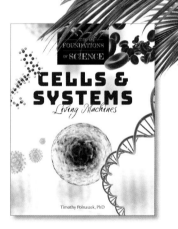

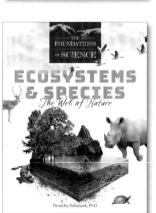
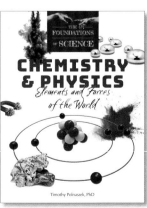